Paul Davies is Professor of Natural Philosophy at the University of Adelaide. He obtained a Ph.D. from the University of London and has held academic appointments at the universities of London, Cambridge and Newcastle upon Tyne. He emigrated to Australia in 1990. His research interests are in the field of black holes, cosmology and quantum gravity, and he has published over one hundred specialist papers as well as several textbooks.

Professor Davies has achieved an international reputation for his ability to explain the significance of advanced scientific ideas in simple language. He is the author of some twenty books, including *Superforce*, *God and the New Physics*, *Other Worlds*, *The Matter Myth* (with John Gribbin), *The Edge of Infinity*, *The Cosmic Blueprint*, *The Last Three Minutes*, *Are We Alone?* and *About Time*. Many of his book are published by Penguin. *The Mind of God* was shortlisted for the 1993 Science Book Prize.

Well known for his media appearances in several countries, Paul Davies has also written and presented a number of TV and radio programmes, including a major series of documentaries on BBC Radio 3 and a six-part television series entitled *The Big Questions*. In 1995 he won the prestigious Templeton Prize, the world's largest award for intellectual endeavour, for his work on the deeper meaning of science. He is also the recipient of a Glaxo Science Writers' Fellowship, an Advance Australia Award and two Eureka Prizes for his contributions to Australian science.

PAUL DAVIES

THE MIND OF GOD

SCIENCE AND THE SEARCH
FOR ULTIMATE MEANING

PENGUIN BOOKS

PENGUIN BOOKS

Published by the Penguin Group
Penguin Books Ltd, 80 Strand, London WC2R 0RL, England
Penguin Putnam Inc., 375 Hudson Street, New York, New York 10014, USA
Penguin Books Australia Ltd, 250 Camberwell Road, Camberwell, Victoria 3124, Australia
Penguin Books Canada Ltd, 10 Alcorn Avenue, Toronto, Ontario, Canada M4V 3B2
Penguin Books India (P) Ltd, 11 Community Centre, Panchsheel Park, New Delhi – 110 017, India
Penguin Books (NZ) Ltd, Cnr Rosedale and Airborne Roads, Albany, Auckland, New Zealand
Penguin Books (South Africa) (Pty) Ltd, 24 Sturdee Avenue, Rosebank 2196, South Africa

Penguin Books Ltd, Registered Offices: 80 Strand, London WC2R 0RL, England

www.penguin.com

First published in Great Britain by Simon & Schuster Ltd 1992
Published in Penguin Books 1993
15

ISBN-13: 978–0–140–15815–1
ISBN-10: 0–140–15815–4

For Caroline, in recognition of
your own search for truth

If we do discover a complete theory, it should in time be understandable in broad principle by everyone, not just a few scientists. Then we shall all, philosophers, scientists, and just ordinary people, be able to take part in the discussion of why it is that we and the universe exist. If we find the answer to that, it would be the ultimate triumph of human reason—for then we would truly know the mind of God.

Stephen Hawking
Concluding sentence of *A Brief History of Time*

Contents

Contents

Contents

Preface

WHEN I WAS A CHILD I used to infuriate my parents by continually asking "why?" Why can't I go out to play? Because it might rain. Why might it rain? Because the weatherman has said so. Why has he said so? Because there are storms coming up from France. Why are there . . . ? And so on. These relentless interrogations normally ended with a desperate "Because God made it that way, and *that's that!*" My childhood discovery (deployed more out of boredom than philosophical acuteness) that the explanation of a fact or circumstance *itself* demanded an explanation, and that this chain might continue indefinitely, has troubled me ever since. Can the chain of explanation really stop somewhere, with God perhaps, or with some superlaw of nature? If so, how does this supreme explanation itself escape the need to be explained? In short, can "that" ever be "that"?

When I became a university student I reveled in the ability of science to provide such breathtaking answers to our questions about the world. The power of science to explain things is so dazzling I found it easy to believe that, given the resources, all the secrets of the universe might be revealed. Yet the "why, why, why . . . ?" worry returned. What lies at the bottom of this magnificent explanatory scheme? What holds it all up? Is there an ultimate level, and if so where did *that* come from? Could one be satisfied with a "that's-that" explanation?

In later years I began doing research on topics like the origin of the universe, the nature of time, and the unification of the laws of physics, and I found myself trespassing on territory that for centuries had been the near-exclusive province of religion. Yet here was science either providing answers to what had been left as dark mysteries, or else discovering that the very concepts from which those mysteries drew their power were actually meaningless or even wrong. My book *God and the New Physics* was a first effort to grapple with this clash of ideologies. *The Mind of God* is a more considered attempt.

Since publication of the first book, a lot of new ideas have emerged at the forefront of fundamental physics: the superstring theory and other approaches to so-called Theories of Everything, quantum cosmology as a means of explaining how the universe might appear from nothing, Stephen Hawking's work on "imaginary time" and the cosmological initial conditions, chaos theory and the concept of self-organizing systems, and advances in the theory of computation and complexity. In addition, there has been an enormous resurgence of interest in what might be crudely described as the science-religion interface. This has taken two distinct forms. First, a greatly increased dialogue between scientists, philosophers, and theologians about the concept of creation and related issues. Second, a growing fashion for mystical thinking and Eastern philosophy, which some commentators have claimed makes deep and meaningful contact with fundamental physics.

I should like to make my own position clear at the outset. As a professional scientist I am fully committed to the scientific method of investigating the world. I believe that science is an immensely powerful procedure for helping us to understand the complex universe in which we live. History has shown that its successes are legion, and scarcely a week passes without some new progress being made. The attraction of the scientific method goes beyond its enormous power and scope, however. There is also its uncompromising honesty. Every new discovery, every theory is required to pass rigorous tests of approval by the scientific community before it is accepted. Of course, in practice, scientists do not always follow the textbook strategies. Sometimes the data are muddled and ambiguous. Sometimes influential scientists sustain dubious theories long after they have been discredited. Occasionally scientists cheat. But these are aberrations. Generally, science leads us in the direction of reliable knowledge.

I have always wanted to believe that science can explain everything, at least in principle. Many nonscientists would deny such a claim resolutely. Most religions demand belief in at least some supernatural events, which are by definition impossible to reconcile with science. I would rather not believe in supernatural events personally. Although I obviously can't prove that they never happen, I see no reason to suppose that they do. My inclination is to assume that the laws of nature are obeyed at all times. But even if one rules out supernatural events, it is still not clear that science could in principle explain everything in the physical universe. There remains that old problem about the end of the explanatory chain. However successful our scientific explanations may be, they always have certain starting assumptions built in. For example, an explanation of some phenomenon in terms of physics presupposes the validity of the laws of physics, which are taken as given. But one can ask where these laws come from in the first place. One could even question the origin of the logic upon which all scientific reasoning is founded. Sooner or later we all have to accept something as given, whether it is God, or logic, or a set of laws, or some other foundation for existence. Thus "ultimate" questions will always lie beyond the scope of empirical science as it is usually defined. So does this mean that the really deep questions of existence are unanswerable? I notice, on perusing the list of my chapter and section titles, that an awful lot of them are questions. At first I thought this was stylistic ineptitude, but I now realize that it reflects my instinctive belief that it is probably impossible for poor old *Homo sapiens* to "get to the bottom of it all." Probably there must always be some "mystery at the end of the universe." But it seems worth pursuing the path of rational inquiry to its limit. Even a proof that the chain of inference is uncompletable would be worth knowing. As we shall see, something of that sort has already been demonstrated in mathematics.

Many practicing scientists are also religious. Following the publication of *God and the New Physics*, I was astonished to discover how many of my close scientific colleagues practice a conventional religion. In some cases they manage to keep these two aspects of their lives separate, as if science rules six days a week, and religion on Sunday. A few scientists, however, make strenuous and sincere efforts to bring their science and their religion into harmony. Usually this entails taking a very liberal view of religious doctrine on the one hand, and on the other hand imbuing the world of physical phenomena with a significance that many of their fellow scientists find unappealing.

Among those scientists who are not religious in a conventional sense, many confess to a vague feeling that there is "something" beyond the surface reality of daily experience, some meaning behind existence. Even hard-nosed atheists frequently have what has been called a sense of reverence for nature, a fascination and respect for its depth and beauty and subtlety, that is akin to religious awe. Indeed, scientists are very emotional people in these matters. There is no greater misconception about scientists than the widespread belief that they are cold, hard, soulless individuals.

I belong to the group of scientists who do not subscribe to a conventional religion but nevertheless deny that the universe is a purposeless accident. Through my scientific work I have come to believe more and more strongly that the physical universe is put together with an ingenuity so astonishing that I cannot accept it merely as a brute fact. There must, it seems to me, be a deeper level of explanation. Whether one wishes to call that deeper level "God" is a matter of taste and definition. Furthermore, I have come to the point of view that mind— i.e., conscious awareness of the world—is not a meaningless and incidental quirk of nature, but an absolutely fundamental facet of reality. That is not to say that *we* are the purpose for which the universe exists. Far from it. I do, however, believe that we human beings are built into the scheme of things in a very basic way.

In what follows I shall attempt to convey the reasons for these beliefs. I shall also examine some of the theories and beliefs of other scientists and theologians, not all of which concur with my own. Much of the discussion involves new advances at the frontiers of science, some of which have led to interesting and exciting ideas about God, creation, and the nature of reality. This book is not, however, intended to be an exhaustive survey of the science-religion interface, but more of a personal quest for understanding. The book is aimed at the general reader, so I have tried to keep the technical aspects to a minimum. No previous knowledge of mathematics or physics is necessary. Some sections, especially chapter 7, involve rather convoluted philosophical arguments, but the reader can pass quickly over these sections without serious problem.

So many people have helped me in this quest, it is impossible to acknowledge them all personally. I have gained much of value from coffee-time conversations with my immediate colleagues at the Universities of Newcastle upon Tyne and Adelaide. I have also received fascinating insights from conversations with John Barrett, John Barrow, Bernard Carr, Philip Davies, George Ellis, David Hooton, Chris

Isham, John Leslie, Walter Mayerstein, Duncan Steel, Arthur Peacocke, Roger Penrose, Martin Rees, Russell Stannard, and Bill Stoeger, and I have been inspired by listening to the lectures of many others. In addition, Graham Nerlich and Keith Ward kindly provided detailed and very valuable comments on certain parts of the manuscript.

Finally, I should like to comment on some matters of terminology. When discussing God, it is often impossible to avoid some sort of personal pronoun. I adhere to the usual convention of using "he." This should not be taken to imply that I believe in a masculine God, or even in the notion of God as a person in any simple sense. Similarly, the use of the word "Man" in the last section refers to the species *Homo sapiens*, not to male persons. When discussing large or small numbers, I use the standard powers-of-ten notation. Thus 10^{20}, for example, means "one followed by twenty zeros," while 10^{-20} stands for $1/10^{20}$.

1

Reason and Belief

HUMAN BEINGS have all sorts of beliefs. The way in which they arrive at them varies from reasoned argument to blind faith. Some beliefs are based on personal experience, others on education, and others on indoctrination. Many beliefs are no doubt innate: we are born with them as a result of evolutionary factors. Some beliefs we feel we can justify, others we hold because of "gut feelings."

Obviously many of our beliefs are wrong, either because they are incoherent, or because they conflict with other beliefs, or with the facts. Two and a half thousand years ago, in ancient Greece, the first systematic attempt was made to establish some sort of common grounds for belief. The Greek philosophers sought a means to formalize human reasoning by providing unassailable rules of logical deduction. By adhering to agreed procedures of rational argument, these philosophers hoped to remove the muddle, misunderstanding, and dispute that so characterize human affairs. The ultimate goal of this scheme was to arrive at a set of assumptions, or axioms, which all reasonable men and women would accept, and from which the resolution of all conflicts would flow.

It has to be said that this goal has never been attained, even if it were possible. The modern world is plagued by a greater diversity of beliefs than ever, many of them eccentric or even dangerous, and rational argument is regarded by a lot of ordinary people as pointless sophistry. Only in science, and especially mathematics, have the ideals of the Greek philosophers been upheld (and in philosophy itself, of course).

When it comes to addressing the really deep issues of existence, such as the origin and meaning of the universe, the place of human beings in the world, and the structure and organization of nature, there is a strong temptation to retreat into unreasoned belief. Even scientists are not immune from this. Yet there is a long and respectable history of attempts to confront such issues by rational and dispassionate analysis. Just how far can reasoned argument take us? Can we really hope to answer the ultimate questions of existence through science and rational inquiry, or will we always encounter impenetrable mystery at some stage? And just what is human rationality anyway?

The Scientific Miracle

Throughout the ages all cultures have extolled the beauty, majesty, and ingenuity of the physical universe. It is only the modern scientific culture, however, that has made any systematic attempt to study the nature of the universe and our place within it. The success of the scientific method at unlocking the secrets of nature is so dazzling it can blind us to the greatest scientific miracle of all: *science works.* Scientists themselves normally take it for granted that we live in a rational, ordered cosmos subject to precise laws that can be uncovered by human reasoning. Yet why this should be so remains a tantalizing mystery. Why should human beings have the ability to discover and understand the principles on which the universe runs?

In recent years more and more scientists and philosophers have begun to study this puzzle. Is our success in explaining the world using science and mathematics just a lucky fluke, or is it inevitable that biological organisms that have emerged from the cosmic order should reflect that order in their cognitive capabilities? Is the spectacular progress of our science just an incidental quirk of history, or does it point to a deep and meaningful resonance between the human mind and the underlying organization of the natural world?

Four hundred years ago science came into conflict with religion because it seemed to threaten Mankind's cozy place within a purpose-built cosmos designed by God. The revolution begun by Copernicus and finished by Darwin had the effect of marginalizing, even trivializing, human beings. People were no longer cast at the center of the great scheme, but were relegated to an incidental and seemingly pointless role in an indifferent cosmic drama, like unscripted extras that have accidentally stumbled onto a vast movie set. This existentialist ethos—that there is no significance in human life beyond what humans themselves invest in it—has become the leitmotif of science. It is for this reason that ordinary people see science as threatening and debasing: it has alienated them from the universe in which they live.

In the chapters that follow I shall present a completely different view of science. Far from exposing human beings as incidental products of blind physical forces, science suggests that the existence of conscious organisms is a *fundamental* feature of the universe. We have been written into the laws of nature in a deep and, I believe, meaningful way. Nor do I regard science as an alienating activity. Far from it. Science is a noble and enriching quest that helps us to make sense of the world in an objective and methodical manner. It does not deny a meaning behind existence. On the contrary. As I have stressed, the fact that science works, and works so well, points to something profoundly significant about the organization of the cosmos. Any attempt to understand the nature of reality and the place of human beings in the universe must proceed from a sound scientific base. Science is not, of course, the only scheme of thought to command our attention. Religion flourishes even in our so-called scientific age. But as Einstein once remarked, religion without science is lame.

The scientific quest is a journey into the unknown. Each advance brings new and unexpected discoveries, and challenges our minds with unusual and sometimes difficult concepts. But through it all runs the familiar thread of rationality and order. We shall see that this cosmic order is underpinned by definite mathematical laws that interweave each other to form a subtle and harmonious unity. The laws are possessed of an elegant simplicity, and have often commended themselves to scientists on grounds of beauty alone. Yet these same simple laws permit matter and energy to self-organize into an enormous variety of complex states, including those that have the quality of consciousness, and can in turn reflect upon the very cosmic order that has produced them.

Among the more ambitious goals of such reflection is the possibility that we might be able to formulate a "Theory of Everything"—a complete description of the world in terms of a closed system of logical truths. The search for such a TOE has become something of a holy grail for physicists. And the idea is undoubtedly beguiling. After all, if the universe is a manifestation of rational order, then we might be able to deduce the nature of the world from "pure thought" alone, without the need for observation or experiment. Most scientists reject this philosophy utterly, of course, hailing the empirical route to knowledge as the only dependable path. But as we shall see, the demands of rationality and logic certainly do impose at least some restrictions on the sort of world that we can know. On the other hand, that same logical structure contains within itself its own paradoxical limitations that ensure we can never grasp the totality of existence from deduction alone.

History has thrown up many physical images for the underlying rational order of the world: the universe as a manifestation of perfect geometrical forms, as a living organism, as a vast clockwork mechanism, and, most recently, as a gigantic computer. All of these images capture some key aspect of reality, though each is incomplete on its own. We shall examine some of the latest thinking about these metaphors, and the nature of the mathematics that describes them. This will lead us to confront the questions: What is mathematics? And why does it work so well in describing the laws of nature? And where do these laws come from anyway? In many cases the ideas are easy to describe; sometimes they are rather technical and abstract. The reader is invited to share this scientific excursion into the unknown, in search of the ultimate basis of reality. Though the going gets rough here and there, and the destination remains shrouded in mystery, I hope that the journey itself will prove exhilarating.

Human Reason and Common Sense

It is often said that the factor which most distinguishes human beings from other animals is our power to reason. Many other animals seem to be aware of the physical world to a greater or lesser extent, and to respond to it, but humans claim more than mere awareness. We also possess some sort of *understanding* of the world, and of our place within it. We are capable of predicting events and of manipulating natural processes to our own ends, and although we are part of the natural world, we somehow distinguish between ourselves and the rest of the physical universe.

In primitive cultures, understanding of the world was limited to everyday affairs, such as the passage of the seasons, or the motion of a slingshot or an arrow. It was entirely pragmatic, and had no theoretical basis, except in magical terms. Today, in the age of science, our understanding has vastly expanded, so that we need to divide knowledge up into distinct subjects—astronomy, physics, chemistry, geology, psychology, and so on. This dramatic progress has come about almost entirely as a result of "the scientific method": experiment, observation, deduction, hypothesis, falsification. The details need not concern us here. What is important is that science demands rigorous standards of procedure and discussion that set reason over irrational belief.

The concept of human reasoning is itself a curious one. We are persuaded by "reasonable" arguments, and feel happiest with those that appeal to "common sense." Yet the processes of human thought are not God-given. They have their origin in the structure of the human brain, and the tasks it has evolved to perform. The operation of the brain, in turn, depends on the laws of physics and the nature of the physical world we inhabit. What we call common sense is the product of thought patterns deeply embedded in the human psyche, presumably because they confer certain advantages in dealing with everyday situations, like avoiding falling objects and hiding from predators. Some aspects of human thought will be fixed by the wiring of our brains, others inherited as "genetic software" from our ancestors of long ago.

The philosopher Immanuel Kant argued that not all our categories of thought derive from sensory experience of the world. He believed that some concepts are *a priori*, by which he meant that, although these concepts are not *necessary truths* in the strictly logical sense, nevertheless all thought would be impossible without them: they would be "necessary for thought." One example Kant gave was our intuitive understanding of three-dimensional space through the rules of Euclidean geometry. He supposed that we are born with this knowledge. Unfortunately, scientists have now discovered that Euclidean geometry is actually wrong! Today, scientists and philosophers generally suppose that even the most basic aspects of human thought must ultimately refer back to observations of the physical world. Probably the concepts that are most deeply etched in our psyche, the things that we find it hard to imagine could be otherwise—such as "common sense" and human rationality—are those that are genetically programmed at a very deep level in our brains.

It is interesting to speculate whether alien beings who evolved under very different circumstances would share our concept of common sense, or indeed any of our thought patterns. If, as some science-fiction writers have mused, life existed on the surface of a neutron star, one could not begin to guess how such beings would perceive and think about the world. It is possible that an alien's concept of rationality would differ from ours so greatly that this being would not be at all persuaded by what we would regard as a rational argument.

Does this mean that human reasoning is suspect? Are we being excessively chauvinistic or parochial in supposing that we can successfully apply the thought patterns of *Homo sapiens* to the great issues of existence? Not necessarily. Our mental processes have evolved as they have precisely because they reflect something of the nature of the physical world we inhabit. What is a surprise is that human reasoning is so successful in framing an understanding of those parts of the world our perceptions can't directly reach. It may be no surprise that human minds can deduce the laws of falling objects, because the brain has evolved to devise strategies for dodging them. But do we have any right to expect extensions of such reasoning to work when it comes to nuclear physics, or astrophysics, for example? The fact that it does work, and works "unreasonably" well, is one of the great mysteries of the universe that I shall be investigating in this book.

But now another issue presents itself. If human reasoning reflects something of the structure of the physical world, would it be true to say that the world is a manifestation of reason? We use the word "rational" to mean "in conformity with reason," so my question is whether, or to what extent, the world is rational. Science is founded on the hope that the world is rational in all its observable aspects. It is possible that there may be some facets of reality which lie beyond the power of human reasoning. This doesn't mean that these facets are necessarily irrational in the absolute sense. Denizens of neutron stars (or supercomputers) might understand things that we, by the very nature of our brains, cannot. So we have to be aware of the possibility that there may be some things with explanations that we could never grasp, and maybe others with no explanation at all.

In this book I shall take the optimistic view that human reasoning is generally reliable. It is a fact of life that people hold beliefs, especially in the field of religion, which might be regarded as irrational. That they are held irrationally doesn't mean they are wrong. Perhaps there is a route to knowledge (such as through mysticism or revelation) that bypasses or transcends human reason? As a scientist I would rather try to take human reasoning as far as it will go. In exploring the frontiers of reason and rationality we will certainly encounter mystery and uncertainty, and in all probability at some stage reasoning will fail us and have to be replaced either by irrational belief or frank agnosticism.

If the world is rational, at least in large measure, what is the origin of that rationality? It cannot arise solely in our own minds, because our minds merely reflect what is already there. Should we seek explanation in a rational Designer? Or can rationality "create itself" by the sheer force of its own "reasonableness"? Alternatively, could it be that on some "larger scale" the world is irrational, but that we find ourselves inhabiting an oasis of apparent rationality, because that is the only "place" where conscious, reasoning beings could find themselves? To explore these sorts of questions further, let us take a more careful look at the different types of reasoning.

Thoughts About Thought

Two sorts of reasoning serve us well, and it is important to keep a clear distinction between them. The first is called "deduction." This is based on the strict rules of logic. According to standard logic, certain statements, such as "A dog is a dog" and "Everything either is, or is not, a dog," are accepted as true, while others, like "A dog is not a dog," are deemed false. A deductive argument starts out with a set of assumptions called "premises." These are statements or conditions which are held to be the case without further questioning, for the purposes of the argument. Obviously the premises should be mutually consistent.

It is widely believed that the conclusion of a logico-deductive argument contains no more than was present in the original premises, so that such an argument can never be used to prove anything genuinely new. Consider, for example, the deductive sequence (known as a "syllogism"):

1. All bachelors are men.
2. Alex is a bachelor.
3. Therefore, Alex is a man.

Statement 3 tells us no more than was present in statements 1 and 2 combined. So, according to this view, deductive reasoning is really only a way of processing facts or concepts so as to present them in a more interesting or useful form.

When deductive logic is applied to a complex set of concepts, the conclusions can often be surprising or unexpected, even if they are merely the outworking of the original premises. A good example is provided by the subject of geometry, which is founded on a collection of assumptions, known as "axioms," on which the elaborate edifice of geometrical theory is erected. In the third century B.C. the Greek geometer Euclid enumerated five axioms on which conventional school geometry is founded, including such things as "Through every two points there is a unique straight line." From these axioms, deductive logic can be used to derive all the theorems of geometry that we learn at school. One of these is Pythagoras' theorem, which, although it has no greater information content than Euclid's axioms, from which it is derived, is certainly not intuitively obvious.

Clearly a deductive argument is only as good as the premises on which it is founded. For example, in the nineteenth century some mathematicians decided to follow up the consequences of dropping Euclid's fifth axiom, which states that through every point it is possible to draw a line parallel to another given line. The resulting "non-Euclidean geometry" turned out to be of great use in science. In fact, Einstein employed it in his general theory of relativity (a theory of gravitation), and, as mentioned, we now know that Euclid's geometry is actually wrong in the real world: crudely speaking, space is curved by gravity. Euclidean geometry is still taught in schools because it remains a very good approximation under most circumstances. The lesson of this story, however, is that it is unwise to consider any axioms as so self-evidently correct that they could not possibly be otherwise.

It is generally agreed that logico-deductive arguments constitute the most secure form of reasoning, though I should mention that even the use of standard logic has been questioned by some. In so-called quantum logic, the rule that something cannot both be and not be such-and-such is dropped. The motivation for this is that in quantum physics the notion of "to be" is more subtle than in everyday experience: physical systems can exist in superpositions of alternative states.

Another form of reasoning that we all employ is called "inductive." Like deduction, induction starts out from a set of given facts or assumptions, and arrives at conclusions from them, but it does so by a process of generalization rather than sequential argument. The prediction that the sun will rise tomorrow is an example of inductive reasoning based on the fact that the sun has faithfully risen every day so far in our experience. And when I let go of a heavy object, I expect it to fall, on the basis of my previous experiences with the pull of gravity. Scientists employ inductive reasoning when they frame hypotheses based on a limited number of observations or experiments. The laws of physics, for instance, are of this sort. The inverse-square law of electric force has been tested in a number of ways, and always confirmed. We call it a law because, on the basis of induction, we reason that the inverse-square property will always hold. However, the fact that nobody has observed a violation of the inverse-square law does not prove it must be true, in the way that, given the axioms of Euclidean geometry, Pythagoras' theorem must be true. No matter on how many individual occasions the law is confirmed, we can never be absolutely certain that it applies unfailingly. On the basis of induction, we may conclude only that it is *very probable* that the law will hold the next time it is tested.

The philosopher David Hume cautioned against inductive reasoning. That the sun has always been observed to rise on schedule, or that the inverse-square law has always been confirmed, is no guarantee that these things will continue in the future. The belief that they will is based on the assumption that "the course of nature continues always uniformly the same." But what is the justification for this assumption? True, it may be the case that a state of affairs B (e.g., dawn) has invariably been observed to follow A (e.g., dusk), but one should not construe this to imply that B is a *necessary* consequence of A. For in what sense might B *have* to follow A? We can certainly conceive of a world where A occurs but B doesn't: there is no logically necessary connection between A and B. Might there be some other sense of necessity, a sort of natural necessity? Hume and his followers deny that there is any such thing.

It seems we are forced to concede that conclusions arrived at inductively are never absolutely secure in the logical manner of deductive conclusions, even though "common sense" is based on induction. That inductive reasoning is so often successful is a (remarkable) property of the world that one might characterize as the "dependability of nature." We all go through life holding beliefs about the world (such as the inevitability of sunrise) which are inductively derived, and considered to be wholly reasonable, and yet rest not on deductive logic, but on the way the world happens to be. As we shall see, there is no logical reason why the world may not have been otherwise. It could have been chaotic in a way that made inductive generalization impossible.

Modern philosophy has been strongly influenced by the work of Karl Popper, who argues that in practice scientists rarely use inductive reasoning in the way described. When a new discovery is made, scientists tend to work backward to construct hypotheses consistent with that discovery, and then go on to deduce other consequences of those hypotheses that can in turn be experimentally tested. If any one of these predictions turns out to be false, the theory has to be modified or abandoned. The emphasis is thus on falsification, not verification. A powerful theory is one that is highly vulnerable to falsification, and so can be tested in many detailed and specific ways. If the theory passes those tests, our confidence in the theory is reinforced. A theory that is too vague or general, or makes predictions concerning only circumstances beyond our ability to test, is of little value.

In practice, then, human intellectual endeavor does not always proceed through deductive and inductive reasoning. The key to major scientific advances often rests with free-ranging imaginative leaps or inspiration. In such cases an important fact or conjecture springs ready-made into the mind of the inquirer, and only subsequently is justification found in reasoned argument. How inspiration comes about is a mystery that raises many questions. Do ideas have a type of independent existence, so that they are "discovered" from time to time by a receptive mind? Or is inspiration a consequence of normal reasoning which takes place hidden in the subconscious, with the result being delivered to the conscious only when complete? If so, how did such an ability evolve? What biological advantages can such things as mathematical and artistic inspiration confer on humans?

A Rational World

The claim that the world is rational is connected with the fact that it is ordered. Events generally do not happen willy-nilly: they are related in some way. The sun rises on cue because the Earth spins in a regular manner. The fall of a heavy object is connected with its earlier release from a height. And so on. It is this interrelatedness of events that gives us our notion of causation. The window breaks because it is struck by a stone. The oak tree grows because the acorn is planted. The invariable conjunction of causally related events becomes so familiar that we are tempted to ascribe causative potency to material objects themselves: the stone actually brings about the breakage of the window. But this is to attribute to material objects active powers that they do not deserve. All one can really say is that there is a correlation between, say, stones rushing at windows and broken glass. Events that form such sequences are therefore not independent. If we could make a record of all events in some region of space over a period of time, we would notice that the record would be crisscrossed by patterns, these being the "causal linkages." It is the existence of these patterns that is the manifestation of the world's rational order. Without them there would be only chaos.

Closely related to causality is the notion of determinism. In its modern form this is the assumption that events are entirely determined by other, earlier events. Determinism carries the implication that the state of the world at one moment suffices to fix its state at a later moment. And because that later state fixes subsequent states, and so on, the conclusion is drawn that everything which ever happens in the future of the universe is completely determined by its present state. When Isaac Newton proposed his laws of mechanics in the seventeenth century, determinism was automatically built into them. For example, treating the solar system as an isolated system, the positions and velocities of the planets at one moment suffice to determine uniquely (through Newton's laws) their positions and velocities at all subsequent moments. Moreover, Newton's laws contain no directionality in time, so the trick works in reverse: the present state suffices to fix uniquely all past states. In this way we can, for example, predict eclipses in the future, and also retrodict their occurrences in the past.

If the world is strictly deterministic, then all events are locked in a matrix of cause and effect. The past and future are contained in the present, in the sense that the information needed to construct the past and future states of the world are folded into its present state just as rigidly as the information about Pythagoras' theorem is folded into the axioms of Euclidean geometry. The entire cosmos becomes a gigantic machine or clockwork, slavishly following a pathway of change already laid down from the beginning of time. Ilya Prigogine has expressed it more poetically: God is reduced to a mere archivist turning the pages of a cosmic history book already written.[1]

Standing in opposition to determinism is indeterminism, or chance. We might say that an event happened by "pure chance" or "by accident" if it was not obviously determined by anything else. Throwing a die and flipping a coin are familiar examples. But are these cases of genuine indeterminism, or is it merely that the factors and forces that determine their outcome are hidden from us, so that their behavior simply *appears* random to us?

Before this century most scientists would have answered yes to the latter question. They supposed that, at rock bottom, the world was strictly deterministic, and that the appearance of random or chance events was entirely the result of ignorance about the details of the system concerned. If the motion of every atom could be known, they reasoned, then even coin tossing would become predictable. The fact that it is unpredictable in practice is because of our limited information about the world. Random behavior is traced to systems that are highly unstable, and therefore at the mercy of minute fluctuations in the forces that assail them from their environment.

This point of view was largely abandoned in the late 1920s with the discovery of quantum mechanics, which deals with atomic-scale phenomena and has indeterminism built into it at a fundamental level. One expression of this indeterminism is known as Heisenberg's uncertainty principle, after the German quantum physicist Werner Heisenberg. Roughly speaking, this states that all measurable quantities are subject to unpredictable fluctuations, and hence to uncertainty in their values. To quantify this uncertainty, observables are grouped into pairs: position and momentum form a pair, as do energy and time. The principle requires that attempts to reduce the level of uncertainty of one member of the pair serves to increase the uncertainty of the other. Thus an accurate measurement of the position of a particle such as an electron, say, has the effect of making its momentum highly uncertain, and vice versa. Because you need to know both the positions and the momenta of the particles in a system precisely if you want to predict its future states, Heisenberg's uncertainty principle puts paid to the notion that the present determines the future exactly. Of course, this supposes that quantum uncertainty is genuinely intrinsic to nature, and not merely the result of some hidden level of deterministic activity. In recent years a number of key experiments have been performed to test this point, and they have confirmed that uncertainty is indeed inherent in quantum systems. The universe really is indeterministic at its most basic level.

So does this mean that the universe is irrational after all? No, it doesn't. There is a difference between the role of chance in quantum mechanics and the unrestricted chaos of a lawless universe. Although there is generally no certainty about the future states of a quantum system, the relative probabilities of the different possible states are still determined. Thus the betting odds can be given that, say, an atom will be in an excited or a nonexcited state, even if the outcome in a particular instance is unknown. This statistical lawfulness implies that, on a macroscopic scale where quantum effects are usually not noticeable, nature seems to conform to deterministic laws.

The job of the physicist is to uncover the patterns in nature and try to fit them to simple mathematical schemes. The question of *why* there are patterns, and why such mathematical schemes are possible, lies outside the scope of physics, belonging to a subject known as metaphysics.

Metaphysics: Who Needs It?

In Greek philosophy, the term "metaphysics" originally meant "that which comes after physics." It refers to the fact that Aristotle's metaphysics was found, untitled, placed after his treatise on physics. But metaphysics soon came to mean those topics that lie beyond physics (we would today say beyond science) and yet may have a bearing on the nature of scientific inquiry. So metaphysics means the study of topics *about* physics (or science generally), as opposed to the scientific subject itself. Traditional metaphysical problems have included the origin, nature, and purpose of the universe, how the world of appearances presented to our senses relates to its underlying "reality" and order, the relationship between mind and matter, and the existence of free will. Clearly science is deeply involved in such issues, but empirical science alone may not be able to answer them, or any "meaning-of-life" questions.

In the nineteenth century the entire metaphysical enterprise began to falter after being critically called into question by David Hume and Immanuel Kant. These philosophers cast doubt not on any particular metaphysical system as such, but on the very meaningfulness of metaphysics. Hume argued that meaning can be attached only to those ideas that stem directly from our observations of the world, or from deductive schemes such as mathematics. Concepts like "reality," "mind," and "substance," which are purported to lie somehow beyond the entities presented to our senses, Hume dismissed on the grounds that they are unobservable. He also rejected questions concerning the purpose or meaning of the universe, or Mankind's place within it, because he believed that none of these concepts can be intelligibly related to things we can actually observe. This philosophical position is known as "empiricism," because it treats the facts of experience as the foundation for all we can know.

Kant accepted the empiricist's premise that all knowledge begins with our experiences of the world, but he also believed, as I have mentioned, that human beings possess certain innate knowledge that is necessary for any thought to take place at all. There are thus two components that come together in the process of thinking: sense data and *a priori* knowledge. Kant used his theory to explore the limits of what human beings, by the very nature of their powers of observation and reasoning, could ever hope to know. His criticism of metaphysics was that our reasoning can apply only to the realm of experience, to the phenomenal world we actually observe. We have no reason to suppose it can be applied to any hypothetical realm that might lie beyond the world of actual phenomena. In other words, we can apply our reasoning to things-as-we-see-them, but this can tell us nothing about the things-in-themselves. Any attempt to theorize about a "reality" that lies behind the objects of experience is doomed to failure.

Although metaphysical theorizing went out of fashion after this onslaught, a few philosophers and scientists refused to give up speculating about what really lies behind the surface appearances of the phenomenal world. Then, in more recent years, a number of advances in fundamental physics, cosmology, and computing theory began to rekindle a more widespread interest in some of the traditional metaphysical topics. The study of "artificial intelligence" reopened debate about free will and the mind-body problem. The discovery of the big bang triggered speculation about the need for a mechanism to bring the physical universe into being in the first place. Quantum mechanics exposed the subtle way in which observer and observed are interwoven. Chaos theory revealed that the relationship between permanence and change was far from simple.

In addition to these developments, physicists began talking about Theories of Everything—the idea that all physical laws could be unified into a single mathematical scheme. Attention began to focus on the nature of physical law. Why had nature opted for one particular scheme rather than another? Why a mathematical scheme at all? Was there anything special about the scheme we actually observe? Would intelligent observers be able to exist in a universe that was characterized by some other scheme?

The term "metaphysics" came to mean "theories about theories" of physics. Suddenly it was respectable to discuss "classes of laws" instead of the actual laws of our universe. Attention was given to hypothetical universes with properties quite different from our own, in an effort to understand whether there is anything peculiar about our universe. Some theorists contemplated the existence of "laws about laws," which act to "select" the laws of our universe from some wider set. A few were prepared to consider the real existence of other universes with other laws.

In fact, in this sense physicists have long been practicing metaphysics anyway. Part of the job of the mathematical physicist is to examine certain idealized mathematical models that are intended to capture only various narrow aspects of reality, and then often only symbolically. These models play the role of "toy universes" that can be explored in their own right, sometimes for recreation, usually to cast light on the real world by establishing certain common themes among different models. These toy universes often bear the name of their originators. Thus there is the Thirring model, the Sugawara model, the Taub-NUT universe, the maximally extended Kruskal universe, and so on. They commend themselves to theorists because they will normally permit exact mathematical treatment, whereas a more realistic model may be intractable. My own work about ten years ago was largely devoted to exploring quantum effects in model universes with only one instead of three space dimensions. This was done to make the problems easier to study. The idea was that some of the essential features of the one-dimensional model would survive in a more realistic three-dimensional treatment. Nobody suggested that the universe really is one-dimensional. What my colleagues and I were doing was exploring hypothetical universes to uncover information about the properties of certain types of physical laws, properties that might pertain to the actual laws of our universe.

Time and Eternity: The Fundamental Paradox of Existence

> "Eternity is time
> Time, eternity
> To see the two as opposites
> Is Man's perversity"

The Book of Angelus Silesius

"I think, therefore I am." With these famous words the seventeenth-century philosopher René Descartes expressed what he took to be the most primitive statement concerning reality about which any thinking persons could agree. Our own existence is our primary experience. Yet even this unexceptionable claim contains within it the essence of a paradox that obstinately runs through the history of human thought. Thinking is a process. Being is a state. When I think, my mental state changes with time. But the "me" to which the mental state refers remains the same. This is probably the oldest metaphysical problem in the book, and it is one which has resurfaced with a vengeance in modern scientific theory.

Though our own selves constitute our primary experience, we also perceive an external world, and we project onto that world the same paradoxical conjunction of process and being, of the temporal and the atemporal. On the one hand, the world continues to exist; on the other hand, it changes. We recognize constancy not just in our personal identities, but in the persistence of objects and qualities in our environment. We frame notions like "person," "tree," "mountain," "sun." These things need not endure forever, but they have a quasi-permanence that enables us to bestow upon them a distinct identity. However, superimposed on this constant backdrop of being is continual change. Things happen. The present fades into the past, and the future "comes into being": the phenomenon of *be*-coming. What we call "existence" is this paradoxical conjunction of being and becoming.

Men and women, perhaps for psychological reasons, being afraid of their own mortality, have always sought out the most enduring aspects of existence. People come and go, trees grow and die, even mountains gradually erode away, and we now know the sun cannot keep burning forever. Is there anything that is truly and dependably constant? Can one find absolute unchanging being in a world so full of becoming? At one time the heavens were regarded as immutable, the sun and stars enduring from eternity to eternity. But we now know that astronomical bodies, immensely old though they may be, have not always existed, nor will they always continue to do so. Indeed, astronomers have discovered that the entire universe is in a state of gradual evolution.

What, then, is absolutely constant? One is inevitably led away from the material and the physical to the realm of the mystical and the abstract. Concepts like "logic," "number," "soul," and "God" recur throughout history as the firmest ground on which to build a picture of reality that has any hope of permanent dependability. But then the ugly paradox of existence rears up at us. For how can the changing world of experience be rooted in the unchanging world of abstract concepts?

Already, at the dawn of systematic philosophy in ancient Greece, Plato confronted this dichotomy. For Plato, true reality lay in a transcendent world of unchanging, perfect, abstract Ideas, or Forms, a domain of mathematical relationships and fixed geometrical structures. This was the realm of pure being, inaccessible to the senses. The changing world of our direct experience—the world of becoming—he regarded as fleeting, ephemeral, and illusory. The universe of material objects was relegated to a pale shadow or parody of the world of Forms. Plato illustrated the relationship between the two worlds in terms of a metaphor. Imagine being imprisoned in a cave with your back to the light. As objects passed by the entrance to the cave they would cast shadows on the cave wall. These shadows would be imperfect projections of the real forms. He likened the world of our observations to the shadow-world of cave images. Only the immutable world of Ideas was "illuminated by the sun of the intelligible."

Plato invented two gods to have dominion over the two worlds. At the pinnacle of the world of Forms was the Good, an eternal and immutable being, beyond space and time. Locked within the half-real and changing world of material objects and forces was the so-called Demiurge, whose task it was to fashion existing matter into an ordered state, using the Forms as a type of template or blueprint. But, being less than perfect, this fashioned world is continually disintegrating and in need of the creative attentions of the Demiurge. Thus arises the state of flux of the world of our sense impressions. Plato recognized a fundamental tension between being and becoming, between the timeless and eternal Forms and the changing world of experience, but made no serious attempt to reconcile the two. He was content to relegate the latter to a partially illusory status, and regard only the timeless and eternal as of ultimate value.

Aristotle, a student of Plato, rejected the concept of timeless Forms, and constructed instead a picture of the world as a living organism, developing like an embryo toward a definite goal. Thus the cosmos was infused with purpose, and drawn toward its goal by final causes. Living things were ascribed souls to guide them in their purposive activity, but Aristotle regarded these souls as immanent in the organisms themselves, and not transcendent in the Platonic sense. This animistic view of the universe laid stress on *process* through progressive goal-oriented change. Thus it might be supposed that, in contrast to Plato, Aristotle gave primacy to becoming over being. But his world remained a paradoxical conjunction of the two. The ends toward which things evolved did not change, nor did the souls. Moreover, Aristotle's universe, though admitting continuous development, had no beginning in time. It contained objects—the heavenly bodies—that were "ungenerated, imperishable, and eternal," moving forever along fixed and perfect circular orbits.

Meanwhile, in the Middle East, the Judaic world view was based on Yahweh's special covenant with Israel. Here the emphasis was placed on God's revelation through history, as expressed in the historical record of the Old Testament, and represented most obviously in Genesis with the account of God's creation of the universe at a finite moment in the past. And yet the Jewish God was still declared to be transcendent and immutable. Again, no real attempt was made to resolve the inevitable paradox of an immutable God whose purposes nevertheless changed in response to historical developments.

A systematic world view that tackled seriously the paradoxes of time had to await the fifth century A.D., and the work of Saint Augustine of Hippo. Augustine recognized that time was part of the physical universe—part of creation—and so he placed the Creator firmly outside the stream of time. The idea of a timeless Deity did not rest easily within Christian doctrine, however. Special difficulty surrounded the role of Christ: What can it mean for a timeless God to become incarnate and die on the cross at some particular epoch in history? How can divine impassibility be reconciled with divine suffering? The debate was continuing in the thirteenth century, when the works of Aristotle became available in translation in the new universities of Europe. These documents had a major impact. A young friar in Paris, Thomas

Aquinas, set out to combine the Christian religion with the Greek methods of rational philosophy. He conceived of a transcendent God inhabiting a Platonic realm beyond space and time. He then attributed a set of well-defined qualities to God—perfection, simplicity, timelessness, omnipotence, omniscience—and attempted to argue logically for their necessity and consistency after the fashion of geometrical theorems. Though his work was immensely influential, Aquinas and his followers had terrible difficulty relating this abstract, immutable Being to the time-dependent physical universe, and the God of popular religion. These and other problems led to Aquinas' work being condemned by the Bishop of Paris, though he was later exonerated and eventually canonized.

In his book *God and Timelessness*, Nelson Pike concludes, after an exhaustive study: "It is now my suspicion that the doctrine of God's timelessness was introduced into Christian theology because Platonic thought was stylish at the time and because the doctrine appeared to have considerable advantage from the point of view of systematic elegance. Once introduced, it took on a life of its own."[2] The philosopher John O'Donnell arrives at the same conclusion. His book *Trinity and Temporality* addresses the conflict between Platonic timelessness and Christian-Judaic historicity: "I am suggesting that as Christianity came into greater contact with Hellenism . . . it sought to achieve a synthesis which was bound to break down precisely at this point. . . . The gospel, combined with certain Hellenistic presuppositions about the nature of God, led to impasses from which the church has yet to extricate itself."[3] I shall return to these "impasses" in chapter 7.

Medieval Europe witnessed the rise of science, and a completely new way of looking at the world. Scientists such as Roger Bacon and, later, Galileo Galilei stressed the importance of obtaining knowledge through precise, quantitative experiment and observation. They regarded Man and nature as distinct, and experiment was seen as a sort of dialogue with nature, whereby her secrets could be unlocked. Nature's rational order, which itself derived from God, was manifested in definite laws. Here an echo of the immutable, timeless Deity of Plato and Aquinas enters science, in the form of eternal laws, a concept that achieved its most persuasive form with the monumental work of Isaac Newton in the seventeenth century. Newtonian physics distinguishes sharply between states of the world, which change from moment to moment, and laws, which remain unchanging. But here once more the difficulty of reconciling being and becoming resurfaces, for how do we account for a flux of time in a world founded upon timeless laws? This "arrow-of-time" conundrum has plagued physics ever since, and is still the subject of intense debate and research.

No attempt to explain the world, either scientifically or theologically, can be considered successful until it accounts for the paradoxical conjunction of the temporal and the atemporal, of being and becoming. And no subject confronts this paradoxical conjunction more starkly than the origin of the universe.

2

Can the Universe Create Itself?

"Science must provide a mechanism for the universe to come into being."

John Wheeler

WE USUALLY THINK of causes as preceding their effects. It is therefore natural to try and explain the universe by appealing to the situation at earlier cosmic epochs. But even if we could explain the present state of the universe in terms of its state a billion years ago, would we really have achieved anything, except moving the mystery back a billion years? For we would surely want to explain the state a billion years ago in terms of some still earlier state, and so on. Would this chain of cause and effect ever end? The feeling that "something must have started it all off" is deeply ingrained in Western culture. And there is a widespread assumption that this "something" cannot lie within the scope of scientific inquiry; it must be in some sense supernatural. Scientists, so the argument goes, might be very clever at explaining this and that. They might even be able to explain everything within the physical universe. But at some stage in the chain of explanation they will reach an impasse, a point beyond which science cannot penetrate. This point is the creation of the universe as a whole, the ultimate origin of the physical world.

This so-called cosmological argument has in one form or another often been used as evidence for the existence of God. Over the centuries it has been refined and debated by many theologians and philosophers, sometimes with great subtlety. The enigma of the cosmic origin is probably the one area where the atheistic scientist will feel uncomfortable. The conclusion of the cosmological argument was, in my opinion, hard to fault until just a few years ago, at which point a serious attempt was made to explain the origin of the universe within the framework of physics. I should say right at the outset that this particular explanation may be quite wrong. However, I don't think that matters. What is at issue is whether or not some sort of supernatural act is necessary to start the universe off. If a plausible scientific theory can be constructed that will explain the origin of the entire physical universe, then at least we know a scientific explanation is possible, whether or not the current theory is right.

Was There a Creation Event?

All debate about the origin of the universe presupposes that the universe *had* an origin. Most ancient cultures inclined to a view of time in which the world has no beginning, but instead experiences endlessly repeating cycles. It is interesting to trace the provenance of these ideas. Primitive tribes always lived closely attuned to nature, depending for their survival on the rhythm of the seasons and other natural periods. Many generations would pass with little alteration in circumstances, so the idea of unidirectional change or historical progress did not occur to them. Questions about the beginning or the fate of the world lay outside their conception of reality. They were preoccupied instead with myths concerning the rhythmic patterns, and the need to propitiate the gods associated with each cycle to ensure continuing fertility and stability.

The rise of the great early civilizations in China and the Middle East made little difference to this outlook. Stanley Jaki, a Hungarian-born Benedictine priest who holds doctorates in both physics and theology, has made a detailed study of ancient beliefs in cyclic cosmology. He points out that the Chinese dynastic system reflected a general indifference toward historical progression. "Their chronological datings restarted with each new dynasty, a circumstance which suggests that for them the flow of time was not linear, but cyclic. Indeed, all events, political and cultural, represented for the Chinese a periodic pattern, a small replica of the interplay of two basic forces in the cosmos, the Yin and the Yang. . . . Success was to alternate with failure, as was progress with decay."[1]

The Hindu system consisted of cycles within cycles, of immense duration. Four yugas made up a mahayuga of 4.32 million years; a thousand mahayugas formed a kalpa, two kalpas constituted a day of

Brahma; the life cycle of Brahma was one hundred years of Brahma, or 311 trillion years! Jaki likens the Hindu cycles to an inescapable treadmill, the mesmerizing effect of which contributed greatly to what he describes as the despair and despondency of the Hindu culture. Cyclicity and the associated fatalism also permeated the Babylonian, Egyptian, and Mayan cosmologies. Jaki relates the story of the Itza, a well-armed Mayan tribe, who voluntarily relinquished control to a small contingent of Spanish soldiers in 1698, having eighty years previously informed two Spanish missionaries that this date marked the onset of their fateful age.

Greek philosophy too was steeped in the concept of eternal cycles, but, in contrast to the pessimistic despair of the poor Mayas, the Greeks believed that their culture represented the top of the cycle—the very pinnacle of progress. The cyclic nature of time in the Greek system was inherited by the Arabs, who remained custodians of the Greek culture until it was transmitted to Christendom in medieval times. Much of the present world view of European cultures can be traced to the monumental clash which then occurred between Greek philosophy and the Judeo-Christian tradition. It is, of course, fundamental to Judaic and Christian doctrine that God created the universe at some specific moment in the past, and that subsequent events form an unfolding unidirectional sequence. Thus a sense of meaningful historical progression—the Fall, the Covenant, the Incarnation and Resurrection, the Second Coming—pervades these religions, and stands in stark contrast to the Greek notion of the eternal return. In their anxiety to adhere to linear, rather than cyclic, time, the early Church Fathers denounced the cyclic world view of the pagan Greek philosophers, notwithstanding their general admiration for all Greek thinking. Thus we find Thomas Aquinas acknowledging the power of Aristotle's philosophical arguments that the universe must always have existed, but appealing for belief in a cosmic origin on biblical grounds.

A key feature of the Judeo-Christian creation doctrine is that the Creator is entirely separate from and independent of his creation; that is, God's existence does not automatically ensure the existence of the universe, as in some pagan schemes where the physical world emanates from the Creator as an automatic extension of his being. Rather, the universe came into existence at a definite instant in time as an act of deliberate supernatural creation by an already existing being.

Straightforward though this concept of creation may seem, it caused intense doctrinal dispute over the centuries, partly because the old texts are somewhat vague on the matter. The biblical description of Genesis, for example, which drew heavily on earlier creation myths from the Middle East, is long on poetry and short on factual details. No clear indication is given of whether God merely brings order to a primordial chaos, or creates matter and light in a pre-existing void, or does something even more profound. Uncomfortable questions abound. What was God doing before he created the universe? Why did he create it at that moment in time rather than some other? If he had been content to endure for eternity without a universe, what caused him to "make up his mind" and create one?

The bible leaves plenty of room for debate on these issues. And debate there certainly has been. In fact, much Christian doctrine concerning the creation was developed long after Genesis was written, and was influenced as much by Greek as by Judaic thought. Two issues in particular are of interest from the scientific point of view. The first is God's relation to time; the second is his relation to matter.

The principal Western religions all proclaim God to be eternal, but the word "eternal" can have two rather different meanings. On the one hand, it can mean that God has existed for an infinite duration of time in the past and will continue to exist for an infinite duration in the future; or it can mean that God is outside of time altogether. As I mentioned in chapter 1, Saint Augustine opted for the latter when he asserted that God made the world "with time and not in time." By regarding time as *part of* the physical universe, rather than something in which the creation of the universe happens, and placing God right outside it altogether, Augustine neatly avoided the problem of what God was doing before the creation.

This advantage is bought at a price, however. Everybody can see the force of the argument that "something must have started it all off." In the seventeenth century it was fashionable to regard the universe as a gigantic machine that had been set in motion by God. Even today, many people like to believe in God's role as a Prime Mover or First Cause in a cosmic chain of causation. But what does it mean for a God who is outside of time to cause anything? Because of this difficulty, believers in a timeless God prefer to emphasize his role in upholding and sustaining the creation at every moment of its existence. No distinction is drawn between creation and preservation: both are, to God's timeless eyes, one and the same action.

God's relationship to matter has similarly been the subject of doctrinal difficulties. Some creation myths, such as the Babylonian version, paint a picture of the cosmos created out of primordial chaos. ("Cosmos" literally means "order" and "beauty"; the latter aspect survives in the modern word "cosmetic.") According to this view, matter precedes, and is ordered by, a supernatural creative act. A similar picture was espoused in classical Greece: Plato's Demiurge was restricted by having to work with already existing material. It was also the position taken by the Christian Gnostics, who regarded matter as corrupt, and therefore a product of the devil rather than God.

Actually, the blanket use of the word "God" in these discussions can be rather confusing, given the wide variety of theological schemes that have been proposed throughout history. Belief in a divine being who starts the universe off and then "sits back" to watch events unfold, taking no direct part in subsequent affairs, is known as "deism." Here God's nature is captured by the image of the perfect watchmaker, a sort of cosmic engineer, who designs and constructs a vast and elaborate mechanism and then sets it going. In contrast to deism is "theism," belief in a God who is creator of the universe, but who also remains directly involved in the day-to-day running of the world, especially the affairs of human beings, with whom God maintains a continuing personal relationship and guiding role. In both deism and theism a sharp distinction is made between God and the world, between creator and creature. God is regarded as wholly other than, and beyond, the physical universe, although he is still responsible for that universe. In the system known as "pantheism," no such separation is made between God and the physical universe. Thus God is identified with nature itself: everything is part of God, and God is in everything. There is also "panentheism," which resembles pantheism in that the universe is part of God, but in which it is not all of God. One metaphor is that of the universe as God's body.

Finally, a number of scientists have proposed a type of God who evolves within the universe, eventually becoming so powerful he resembles Plato's Demiurge. One can envisage, for example, intelligent life or even machine intelligence gradually becoming more advanced and spreading throughout the cosmos, gaining control over larger and larger portions until its manipulation of matter and energy is so refined that this intelligence would be indistinguishable from nature itself. Such a God-like intelligence could develop from our own descendants, or even have developed already from some extraterrestrial community or communities. Fusion of two or more different intelligences during this evolutionary process is conceivable. Systems of this sort have been proposed by the astronomer Fred Hoyle, the physicist Frank Tipler, and the writer Isaac Asimov. The "God" in these schemes is clearly less than the universe and, though immensely powerful, is not omnipotent, and cannot be regarded as the creator of the universe as a whole, only of part of its organized content. (Unless, that is, some peculiar arrangement of backward causation is introduced, whereby the superintelligence at the end of the universe acts backward in time to create that universe, as part of a self-consistent causal loop. There are hints of this in the ideas of physicist John Wheeler. Fred Hoyle has also discussed such a scheme, but not in the context of an all-embracing creation event.)

Creation from Nothing

The pagan creation myths assume the existence of both material stuff and a divine being, and so are fundamentally dualistic. By contrast, the early Christian Church settled on the doctrine of "creation *ex nihilo*," in which God alone is necessary. He is taken to have created the entire universe from nothing. The origin of all things visible and invisible, including matter, is thus attributed to a free creative act by God. An important component in this doctrine is God's omnipotence: there is no limitation to his creative power, as was the case with the Greek Demiurge. Indeed, not only is God not limited to work with pre-existing matter, he is not limited by pre-existing physical laws either, for part of his creative act was to bring those laws into being and thereby establish the order and harmony of the cosmos. The Gnostic belief that matter is corrupt is rejected as being incompatible with the Incarnation of Christ. On the other hand, neither is matter divine, as in pantheistic schemes, where all nature is infused with God's presence. The physical universe—God's creature—is regarded as distinct and separate from its creator.

The importance of the distinction between creator and creature in this system is that the created world depends absolutely for its existence on the creator. If the physical world itself were divine, or somehow emanated directly from the creator, then it would share the creator's necessary existence. But because it was created from nothing, and because the creative act was a free choice of the creator, the universe does not have to exist. Thus Augustine writes: "You created something, and that something out of nothing. You made heaven and earth, not out of yourself, for then they would have been equal to your Only-begotten, and through this equal also to you."[2] The most obvious distinction between creator and creature is that the creator is eternal whereas the created world has a beginning. Thus the early Christian theologian Iranaeus wrote: "But the things established are distinct from Him who has established them, and what have been made from Him who has made them. For He is Himself uncreated, both without beginning and end, and lacking nothing. He is Himself sufficient for this very thing, existence; but the things which have been made by Him have received a beginning."[3]

Even today there remain doctrinal differences within the main branches of the Church, and still greater differences among the various world religions, concerning the meaning of creation. These range from the ideas of Christian and Islamic fundamentalists, based on a literal interpretation of the traditional texts, to those of radical Christian thinkers who prefer a totally abstract view of creation. But all agree that in one sense or another the physical universe on its own is incomplete. It cannot explain itself. Its existence ultimately demands something outside of itself, and can be understood only from its dependence on some form of divine influence.

The Beginning of Time

Turning to the scientific position on the origin of the universe, one can again ask about the evidence that there actually was an origin. It is certainly possible to conceive of a universe of infinite duration, and for much of the modern scientific era, following the work of Copernicus, Galileo, and Newton, scientists did in fact generally believe in an eternal cosmos. There were, however, some paradoxical aspects to this belief. Newton was worried about the consequences of his law of gravity, which holds that all matter in the universe attracts all other matter. He was puzzled about why the whole universe does not simply fall together into one great mass. How can the stars hang out there in space forever, unsupported, without being pulled toward each other by their mutual gravitational forces? Newton proposed an ingenious solution. For the universe to collapse to its center of gravity, there has to *be* a center of gravity. If, however, the universe were infinite in spatial extent, and on average populated uniformly by stars, then there would be no privileged center toward which the stars could fall. Any given star would be tugged similarly in all directions, and there would be no resultant force in any given direction.

This solution is not really satisfactory, because it is mathematically ambiguous: the various competing forces are all infinite in magnitude. So the mystery of how the universe avoids collapse kept recurring, and persisted into the present century. Even Einstein was perplexed. His own theory of gravitation (the general theory of relativity) was formulated in 1915, and almost immediately "fixed up" in an attempt to explain the stability of the cosmos. The fix consisted of an extra term in his gravitational-field equations corresponding to a force of repulsion—a type of antigravity. If the strength of this repulsive force was tuned to match the gravitational pull of all the cosmic bodies on each other, attraction and repulsion could be balanced to produce a static universe. Alas, the balancing act turned out to be unstable, so that the merest disturbance would cause one or the other of the competing forces to win out, either dispersing the cosmos in a runaway outward rush, or sending it crashing inward.

Nor was the collapsing-cosmos mystery the only problem with an eternal universe. There was also something called Olbers' paradox, which concerned the darkness of the night sky. The difficulty here was that, if the universe is infinite in spatial extent as well as in age, then light from an infinity of stars will be pouring down upon the Earth from the heavens. A simple calculation shows that the sky could not be dark under these circumstances. The paradox can be resolved by assuming a finite age for the universe, for in that case we will be able to see only those stars whose light has had time to travel across space to Earth since the beginning.

Today, we recognize that no star could keep burning forever anyway. It would run out of fuel. This serves to illustrate a very general principle: an eternal universe is incompatible with the continuing existence of irreversible physical processes. If physical systems can undergo irreversible change at a finite rate, then they will have completed those changes an infinite time ago. Consequently we could not be witnessing such changes (such as the production and emission of starlight) now.

In fact, the physical universe abounds with irreversible processes. In some respects it is rather like a clock slowly running down. Just as a clock cannot keep running forever, so the universe cannot have been "running" forever without being "rewound."

These problems began to force themselves on scientists during the mid-nineteenth century. Until then physicists had dealt with laws that are symmetric in time, displaying no favoritism between past and future. Then the investigation of thermodynamic processes changed that for good. At the heart of thermodynamics lies the second law, which forbids heat to flow spontaneously from cold to hot bodies, while allowing it to flow from hot to cold. This law is therefore not reversible: it imprints upon the universe an arrow of time, pointing the way of unidirectional change. Scientists were quick to draw the conclusion that the universe is engaged in a one-way slide toward a state of thermodynamic equilibrium. This tendency toward uniformity, wherein temperatures even out and the universe settles into a stable state, became known as the "heat death." It represents a state of maximum molecular disorder, or entropy. The fact that the universe has not yet so died—that is, it is still in a state of less-than-maximum entropy—implies that it cannot have endured for all eternity.

In the 1920s astronomers discovered that the traditional picture of a static universe was in any case wrong. They found that the universe is, in fact, expanding, with the galaxies rushing away from each other. This is the basis of the well-known big-bang theory, according to which the entire universe came into existence abruptly, about fifteen billion years ago, in a gigantic explosion. The expansion seen today can be regarded as a vestige of that primeval outburst. The discovery of the big bang has often been hailed as confirmation of the biblical account of Genesis. Indeed, in 1951 Pope Pius XII alluded to it in an address to the Pontifical Academy of Sciences. Of course, the big-bang scenario bears only the most superficial resemblance to Genesis, so that the latter has to be interpreted in an almost completely symbolic way for any connection to be made. About the best that can be said is that both accounts demand an abrupt rather than a gradual beginning, or no beginning at all.

The big-bang theory automatically evades the paradoxes of an eternal cosmos. Because the universe is finite in age, there are no problems with irreversible processes. The universe evidently began, in some sense, "wound up," and is currently still busy unwinding. The night sky is dark because we can see only a finite distance into space (about fifteen billion light-years), this being the maximum distance from which light has been able to travel to Earth since the beginning. Nor is there a difficulty about the universe collapsing under its own weight. Because the galaxies are flying apart, they avoid falling together, at least for a while.

Nevertheless, the theory solves one set of problems only to be confronted by another, not least of which is to explain what caused the big bang in the first place. It is here that we encounter an important subtlety about the nature of the big bang. Some popular accounts give the impression that it was the explosion of a concentrated lump of matter located at some particular place in a pre-existing void. This is badly misleading. The big-bang theory is based on Einstein's general theory of relativity. One of the main features of general relativity is that the affairs of matter cannot be separated from the affairs of space and time. It is a linkage that has profound implications for the origin of the universe. If one imagines "running the cosmic movie backward," then the galaxies get closer and closer together until they merge. Then the galactic material gets squeezed more and more until a state of enormous density is reached. One might wonder whether there is any limit to the degree of compression as we pass back to the moment of explosion.

It is easy to see that there can be no simple limit. Suppose that there were a state of maximum compression. This would imply the existence of some sort of outward force to overcome the enormous gravity; otherwise gravity would win, and the material would be still more compressed. Furthermore, this outward force would have to be truly enormous, because the inward force of gravity rises without limit as the compression rises. So what could this stabilizing force be? A type of pressure or material stiffness, perhaps—who knows what forces nature might deploy under such extreme conditions? However, although we do not know the details of the forces, certain general considerations must still apply. For example, as the material gets stiffer and stiffer, so the speed of sound in the cosmic material gets faster. It seems clear that, if the stiffness of the primeval cosmic material were to become large enough, the speed of sound would exceed the speed of light. But this is strictly contrary to the theory of relativity, which requires that no physical influence should travel faster than light. Therefore, the material cannot ever have been infinitely stiff. Consequently, at some stage during the compression, the force of gravity would have been greater than the stiffness force, which implies that the stiffness would have been unable to contain gravity's compressing tendency.

The conclusion that was drawn concerning this tussle between primeval forces was that, under conditions of extreme compression such as occurred during the big bang, there is no force in the universe capable of beating off the crushing power of gravity. The crushing has no limit. If the matter in the universe were spread uniformly, then it must have been infinitely compressed at the first moment. In other words, the entire cosmos would have been squeezed into a single point. At this point the gravitational force, and the density of material, were infinite. A point of infinite compression is known to mathematical physicists as a "singularity."

Although one is led on quite elementary grounds to expect a singularity at the origin of the universe, it required a mathematical investigation of some delicacy to establish the result rigorously. This investigation was mainly the work of British mathematical physicists Roger Penrose and Stephen Hawking. In a series of powerful theorems, they proved that a big-bang singularity is inevitable as long as gravity remains an attractive force under the extreme conditions of the primeval universe. The most significant aspect of their results is that a singularity isn't avoided even if the cosmic material is distributed unevenly. It is a general feature of a universe described by Einstein's theory of gravitation—or, for that matter, any similar theory.

There was a lot of resistance to the idea of a big-bang singularity among physicists and cosmologists when it was first mooted. One reason for this concerns the above-mentioned fact that matter, space, and time are linked in the general theory of relativity. This linkage carries important implications for the nature of the expanding universe. Naïvely, one might suppose that the galaxies are rushing apart through space. A more accurate picture, however, is to envisage space itself as swelling or stretching. That is, the galaxies move apart because the space between them expands. (Readers who are unhappy about the idea that space can stretch are referred to my book *The Edge of Infinity* for further discussion.) Conversely, in the past, space was shrunken. If we consider the moment of infinite compression, space was infinitely shrunk. But if space is infinitely shrunk, it must literally disappear, like a balloon that shrivels to nothing. And the all-important linkage of space, time, and matter further implies that time must disappear too. There can be no time without space. Thus the material singularity is also a space-time singularity. Because all our laws of physics are formulated in terms of space and time, these laws cannot apply beyond the point at which space and time cease to exist. Hence the laws of physics must break down at the singularity.

The picture that we then obtain for the origin of the universe is a remarkable one. At some finite instant in the past the universe of space, time, and matter is bounded by a space-time singularity. The coming-into-being of the universe is therefore represented not only by the abrupt appearance of matter, but of space and time as well.

The significance of this result cannot be overstressed. People often ask: Where did the big bang occur? The bang did not occur at a point in space at all. Space itself came into existence with the big bang. There is similar difficulty over the question: What happened before the big bang? The answer is, there was no "before." Time itself began at the big bang. As we have seen, Saint Augustine long ago proclaimed that the world was made with time and not in time, and that is precisely the modern scientific position.

Not all scientists were prepared to go along with this, however. While accepting the expansion of the universe, some cosmologists attempted to construct theories that nevertheless avoided a singular origin to space and time.

Cyclic World Revisited

In spite of the strong Western tradition for a created universe and a linear time, the lure of the eternal return always lies just beneath the surface. Even during the modern big-bang era there have been attempts to reinstate a cyclic cosmology. As we have seen, when Einstein formulated his general theory of relativity, scientists still believed in a static cosmos, and this prompted Einstein to "fix up" his equations to create a gravitational-levitational equilibrium. Meanwhile, however, an obscure Russian meteorologist by the name of Alexander Friedmann began studying Einstein's equations and their implications for cosmology. He discovered several interesting solutions, all of which describe a universe that either expands or contracts. One set of solutions corresponds to a universe that starts out at a big bang, expands at an ever-diminishing rate, and then starts to contract again. The contracting phase mirrors the expanding phase, so that the contraction gets faster and faster until the universe disappears at a "big crunch"—a catastrophic implosion like the big bang in reverse. This cycle of expansion and contraction can then be continued into another cycle, then another, and so on *ad infinitum* (see figure 1). In 1922 Friedmann sent the details of his periodic-universe model to Einstein, who wasn't impressed. It was only some years later, with the discovery by Edwin Hubble and other astronomers that the universe is indeed expanding, that Friedmann's work came to be properly recognized.

Friedmann's solutions do not compel the universe to oscillate with phases of expansion and contraction. They also provide for a universe which starts out at a big bang and goes on expanding forever. Which of these alternatives prevails turns out to depend on the amount of matter that exists in the universe. Basically, if there is enough matter, its gravity will eventually halt the cosmic dispersal, and bring about recollapse. Thus Newton's fear of cosmic collapse would actually be realized, albeit only after billions of years had elapsed. Measurements reveal that the stars constitute only about 1 percent of the density needed to collapse the universe. However, there is strong evidence for a large amount of dark or invisible matter, possibly enough to make up the deficit. Nobody is sure what this "missing matter" is.

If there is sufficient matter to cause recontraction, we have to consider the possibility that the universe might be pulsating, as indicated in figure 1. Many popular books on cosmology feature the pulsating model, and point out its consistency with Hindu and other Eastern cosmologies of a cyclic nature. Could it be that Friedmann's oscillating solution is the scientific counterpart of the ancient idea of the eternal return, and that the multibillion-year duration from big bang to big crunch represents the Great Year of the Life Cycle of Brahma?

Appealing though these analogies may seem, they fail to hold up to scrutiny. First of all, the model isn't strictly periodic in the mathe-

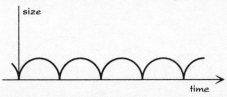

FIGURE 1. Oscillating universe. The graph shows how the size of the universe varies with time as it expands and contracts in a cyclic manner.

matical sense. The points of turnaround from big crunch to big bang are actually singularities, which means that the equations concerned break down there. In order for the universe to bounce from a contraction to an expansion without encountering singularities, it is necessary for something to reverse the pull of gravity and propel the material outward again. In essence, a bounce is possible only if the motion of the universe is overwhelmed by a huge repulsive (i.e., levitational) force, such as the "fix-up" force Einstein suggested, but bigger in magnitude by an enormous factor.

Even if a mechanism to do this could be contrived, the cyclicity of the model concerns only the gross motion of the cosmos, and ignores the physical processes within it. The second law of thermodynamics still demands that these processes generate entropy, and that the total entropy of the universe goes on rising from one cycle to the next. The result is a rather curious effect, discovered by Richard Tolman in the 1930s. Tolman found that, as the entropy of the universe rises, so the cycles grow bigger and bigger, and last longer and longer (figure 2). The upshot is that the universe isn't strictly cyclic at all. Strangely, in spite of the continued rise in entropy, the universe never reaches thermodynamic equilibrium—there is no maximum-entropy state. It just goes on pulsating forever, generating more and more entropy along the way.

In the 1960s the astronomer Thomas Gold believed he had found a truly cyclic model of the universe. Gold knew that an eternally static universe is untenable because it would reach thermodynamic equilibrium in a finite time. He was struck by the fact that the expansion of the universe worked against thermodynamic equilibrium by continuously cooling the cosmic material (this is the familiar principle that matter cools when expanded). It seemed to Gold that the rise in cosmic entropy could be attributed to the fact that the universe is expanding. But this conclusion carried with it the hint of a remarkable prediction:

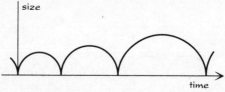

FIGURE 2. In a more realistic model of an oscillating universe, the cycles grow larger with time.

if the universe were to contract, everything would run backward—entropy would fall again, and the second law of thermodynamics would be reversed. In a sense, time would flow backward. Gold pointed out that this reversal would apply to all systems, including the human brain and memory, so that the psychological arrow of time would also be reversed: we would "remember the future" instead of the past. Any conscious beings living in what we would regard as the contracting phase would reverse our definitions of past and future, and also consider themselves as living during the expanding phase of the universe (figure 3). By their definition, ours would be the contracting phase. If, as a result of the reversal, the universe were truly symmetric in time, then the final state of the universe at the big crunch would be identical to its state at the big bang. These two events could therefore be identified, and time closed into a loop. In that case the universe would indeed be cyclic.

The time-symmetric universe was also investigated by John Wheeler, who conjectured that the turnaround might not occur abruptly, but gradually, like the turning of the tide. Instead of the arrow of time suddenly reversing at the epoch of maximum expansion, perhaps it might slowly falter and then fade away altogether before swinging round to point the other way. Wheeler speculated that as a result some apparently irreversible processes, such as the decay of radioactive nuclei, might show signs of slowing down ahead of the reversal. He suggested that a comparison of the rates of radioactive decay now with their values in the remote past might indicate such a slowing down.

Another phenomenon that displays a distinct arrow of time is the emission of electromagnetic radiation. A radio signal, for example, is

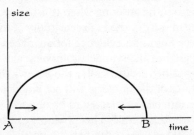

FIGURE 3. Time-reversing universe. During the expanding phase time runs forward, and during the contracting phase it runs backward. As a result, it is possible to identify the first and last moments A and B, thus closing time into a loop.

always received after it is sent, never before. This is because, when radio transmitters generate radio waves, the waves flow outward from the antenna into the depths of the universe. We never observe organized patterns of radio waves coming from the edges of the universe and converging onto radio antennae. (The technical term for outflowing waves is "retarded," whereas inflowing waves are "advanced.") If, however, the arrow of time were to reverse in the contracting phase of the universe, then the direction of radio-wave motion would also have to reverse—retarded waves would be replaced by advanced waves. In the context of Wheeler's "turning of the tide," this would suggest that close to the big bang all radio waves would be retarded; then, as the epoch of maximum expansion was approached, so increasing amounts of advanced waves would occur. At maximum, there would be equal advanced and retarded waves, whereas during the contracting phase advanced waves would dominate. If this idea is correct, it would imply that there is a very slight admixture of advanced radio waves at our present cosmic epoch. In effect, these would be radio waves "from the future."

Fanciful though the idea may seem, it was put to the test in an experiment performed by the astronomer Bruce Partridge in the 1970s. The principle of the experiment is that, if radio waves emitted by an antenna are directed toward a screen where they are absorbed, the waves will be 100 percent retarded; if they are allowed to flow away into space, part of them will continue unaffected until the "tide has turned." The latter, but not the former, set of waves might then possess a tiny advanced component. If that is the case, the advanced waves will put back into the antenna a small fraction of the energy that the retarded waves took out. The effect would be to produce a slight difference in the energy drain from the antenna when it is beamed at the screen as opposed to space. In spite of the high sensitivity of the measurements, however, Partridge found no evidence for advanced waves.

Beguiling though the time-symmetric universe may be, it is very hard to argue for it plausibly. Statistically, the overwhelming majority of possible initial states of the universe will *not* produce reversal, so only if the state of the universe is selected to belong to a very peculiar and special set will "the tide turn." The situation can be compared to a bomb exploding inside a steel container: it is possible to imagine all the fragments of the bomb rebounding in unison from the walls of the container and coming back together to reconstitute the bomb. That sort of conspiratorial behavior is not strictly impossible, but it requires an incredibly contrived set of circumstances.

Nevertheless, the time-symmetric universe idea has proved sufficiently compelling that even Stephen Hawking recently flirted with it as part of his quantum-cosmology program, which I shall explain shortly. However, following more detailed investigation, Hawking admitted that his proposal was misconceived.

Continuous Creation

Thomas Gold tells the story that one evening in the late 1940s he and Hermann Bondi were walking back from the cinema, having seen a movie called *Dead of Night,* about dreams within dreams that formed an endless sequence. On the way home it suddenly occurred to them that the theme of the film might be an allegory for the universe. Perhaps there *was* no beginning, no big bang even. Maybe the universe instead has a means of continuously replenishing itself so that it can keep going forever.

In the subsequent months Bondi and Gold fleshed out their idea. The central feature of the Bondi-Gold theory is that there was no big-bang origin of the universe at which all matter was created. Instead, as the universe expands, new particles of matter are continuously created to fill up the gaps so that the average density of matter in the universe remains unchanged. Any individual galaxy will pass through a life cycle of evolution, culminating in its death when the stars burn out, but new galaxies are able to form from the newly created matter. At any given time there will be a mixture of galaxies of various ages, but the very old galaxies will be very sparsely distributed, because the universe will have expanded a lot since their birth. Bondi and Gold envisaged the rate of expansion of the universe to remain constant, and the rate of matter creation to be just such as to maintain a constant average density. The situation is similar to that of a river which looks the same overall, even though water is continuously flowing through it. The river is not static, but in a steady state. The theory therefore became known as the "steady-state theory" of the universe.

The steady-state universe has no beginning or end, and looks the same on average at all cosmic epochs, in spite of the expansion. The model avoids the heat death, because the injection of new matter also injects negative entropy: to return to the watch analogy, it continuously rewinds the watch. Bondi and Gold gave no detailed mechanism to explain how matter is created, but their colleague Fred Hoyle had been working on just this problem. Hoyle investigated the possibility of a "creation field" that would have the effect of producing new particles of matter. Because matter is a form of energy, Hoyle's mechanism might be construed as violating the law of conservation of energy, but this need not be so. The creation field itself carries negative energy, and by arranging things carefully, it is possible for the positive energy of the created matter to be compensated exactly by the enhanced negative energy of the creation field. From a mathematical study of this interplay, Hoyle discovered that his creation-field cosmological model automatically tended toward, and then remained in, the steady-state condition required by the theory of Bondi and Gold.

Hoyle's work gave the necessary theoretical underpinning to ensure that the steady-state theory was taken seriously, and for a decade or more it was considered to be an equal contender with the big-bang theory. Many scientists, including the originators of the steady-state theory, felt that by abolishing the big bang they had once and for all removed the need for any sort of supernatural explanation for the universe. In a universe with no beginning there is no need for a creation event or a creator, and a universe with a physical creation field to make it "self-winding" doesn't require any divine intervention to keep it running.

Actually, the conclusion is a *non sequitur*. The fact that the universe might have no origin in time does not explain its existence, or why it has the form it does. Certainly it does not explain why nature possesses the relevant fields (such as the creation field) and physical principles that establish the steady-state condition. Ironically, some theologians have actually welcomed the steady-state theory as a *modus operandi* for God's creative activity. After all, a universe that lives forever, avoiding the heat death, has considerable theological appeal. Around the turn of the century the mathematician and philosopher Alfred North Whitehead founded the so-called process school of theology. Process theologians reject the traditional Christian concept of creation out of nothing in favor of a universe that had no beginning. God's creative activity manifests itself instead as an ongoing process, a creative advance in nature's activity. I shall return to the topic of creative cosmology in chapter 7.

In the event, the steady-state theory fell out of favor not on philosophical grounds, but because it was falsified by observations. The theory made the very specific prediction that the universe should look the same on average at all epochs, and the advent of large radio telescopes enabled this prediction to be tested. When astronomers observe very distant objects, these do not appear as they are today, but as they were in the remote past, when the light or radio waves left them on their long journey to Earth. These days, astronomers can study objects that are many billions of light-years away, so that we see them as they were many billions of years ago. Thus a deep space survey can provide "snapshots" of the universe at successive epochs, for comparison. By the mid-1960s it became clear that several billion years ago the universe would have looked very different from the way it does now, in particular vis-à-vis the numbers of various types of galaxies.

The final nail in the coffin of the steady-state theory came in 1965 with the discovery that the universe is bathed in heat radiation at a temperature of about three degrees above absolute zero. This radiation is believed to be a direct relic of the big bang, a sort of fading glow of the primeval heat that accompanied the birth of the cosmos. It is hard to understand how such a radiation bath could have arisen without the cosmic material having once been highly compressed and exceedingly hot. Such a state does not occur in the steady-state theory. Of course, the fact that the universe is not in a steady state does not mean that continuous creation of matter is impossible, but the motivation for Hoyle's creation field is largely undermined once it is established that the universe is evolving. Nearly all cosmologists now accept that we live in a universe that had a definite beginning in a big bang, and is developing toward an uncertain end.

If one accepts the idea that space, time, and matter had their origin in a singularity that represents an absolute boundary to the physical universe in the past, a number of puzzles follow. There is still the problem of what caused the big bang. However, this question must now be seen in a new light, for it is not possible to attribute the big bang to anything that happened *before* it, as is usually the case in discussions of causation. Does this mean the big bang was an event without a cause? If the laws of physics break down at the singularity, there can be no explanation in terms of those laws. Therefore, if one insists on a reason for the big bang, then this reason must lie beyond physics.

Did God Cause the Big Bang?

Many people have an image of God as a sort of pyrotechnic engineer, lighting the blue touch-paper to ignite the big bang, and then sitting back to watch the show. Unfortunately, this simple picture, while highly compelling to some, makes little sense. As we have seen, a supernatural creation cannot be a causative act in time, for the coming-into-being of time is part of what we are trying to explain. If God is invoked as an explanation for the physical universe, then this explanation cannot be in terms of familiar cause and effect.

This recurring problem of time was recently addressed by the British physicist Russell Stannard, who draws the analogy between God and the author of a book. A completed book exists in its entirety, although we humans read it in a time sequence from the beginning to the end. "Just as an author does not write the first chapter, and then leave the others to write themselves, so God's creativity is not to seem as uniquely confined to, or even especially invested in, the event of the Big Bang. Rather, his creativity has to be seen as permeating equally all space and all time: his role as Creator and Sustainer merge."[4]

Quite apart from the problems of time, there are several additional pitfalls involved in invoking God as an explanation for the big bang. To illustrate them I shall relate an imaginary conversation between a theist (or, more properly, a deist), who claims that God created the universe, and an atheist, who "has no need of this hypothesis."

ATHEIST: At one time, gods were used as an explanation for all sorts of physical phenomena, such as the wind and the rain and the motion of the planets. As science progressed, so supernatural agents were found to be superfluous as an explanation for natural events. Why do you insist on invoking God to explain the big bang?

THEIST: Your science cannot explain everything. The world is full of mystery. For example, even the most optimistic biologists admit that they are baffled by the origin of life.

ATHEIST: I agree that science hasn't explained everything, but that doesn't mean it can't. Theists have always been tempted to seize on any process that science could not at the time explain and claim that God was still needed to explain it. Then, as science progressed, God got squeezed out. You should learn the lesson that this "God of the gaps" is an unreliable hypothesis. As time goes on, there are fewer and fewer gaps for him to inhabit. I personally see no problem in science explaining all natural phenomena, including the origin of life. I concede that the origin of the universe is a tougher nut to crack. But if, as it seems, we have now reached the stage where the only remaining gap is the big bang, it is highly unsatisfying to invoke the concept of a supernatural being who has been displaced from all else, in this "last-ditch" capacity.

THEIST: I don't see why. Even if you reject the idea that God can act directly in the physical world once it has been created, the problem of the ultimate origin of that world is in a different category altogether from the problem of explaining natural phenomena once that world exists.

ATHEIST: But unless you have other reasons to believe in God's existence, then merely proclaiming "God created the universe" is totally *ad hoc*. It is no explanation at all. Indeed, the statement is essentially devoid of meaning, for you are merely defining God to be that agency which creates the universe. My understanding is no further advanced by this device. One mystery (the origin of the universe) is explained only in terms of another (God). As a scientist I appeal to Occam's razor, which then dictates that the God hypothesis be rejected as an unnecessary complication. After all, I am bound to ask, what created God?

THEIST: God needs no creator. He is a necessary being—he must exist. There is no choice in the matter.

ATHEIST: But one might as well assert that the universe needs no creator. Whatever logic is used to justify God's necessary existence could equally well, and with an advantageous gain in simplicity, be applied to the universe.

THEIST: Surely scientists often follow the same reasoning as I have. Why does a body fall? Because gravity acts on it. Why does gravity act on it? Because there is a gravitational field. Why? Because space-time is curved. And so on. You are replacing one description with another, deeper description, the sole purpose of which is to explain the thing you started with, namely, falling bodies. Why do you then object when I invoke God as a deeper and more satisfying explanation of the universe?

ATHEIST: Ah, but that's different! A scientific theory should amount to much more than the facts it is trying to explain. Good theories provide a simplifying picture of nature by establishing connections between hitherto disconnected phenomena. Newton's gravitational theory, for example, demonstrated a connection between the ocean tides and the motion of the moon. In addition, good theories suggest observational tests, such as predicting the existence of new phenomena. They also provide detailed mechanistic accounts of precisely how the physical processes of interest happen in terms of the concepts of the theory. In the case of gravitation, this is through a set of equations that connect the strength of the gravitational field with the nature of the gravitating sources. This theory gives you a precise mechanism for how things work. By contrast, a God who is invoked only to explain the big bang fails in all three criteria. Far from simplifying our view of the world, a Creator introduces an additional complicating feature, itself without explanation. Second, there is no way we can test the hypothesis experimentally. There is only one place where such a God is manifested—namely, the big bang—and that is over and done with. Finally, the bald statement "God created the universe" fails to provide any real explanation unless it is accompanied by a detailed mechanism. One wants to know, for example, what properties to assign this God, and precisely how he goes about creating the universe, why the universe has the form it does, and so on. In short, unless you can either provide evidence in some other way that such a God exists, or else give a detailed account of *how* he made the universe that even an atheist like me would regard as deeper, simpler, and more satisfying, I see no reason to believe in such a being.

THEIST: Nevertheless, your own position is highly unsatisfactory, for you admit that the reason for the big bang lies outside the scope of science. You are forced to accept the origin of the universe as a brute fact, with no deeper level of explanation.

ATHEIST: I would rather accept the existence of the universe as a brute fact than accept God as a brute fact. After all, there has to be a universe for us to be here to argue about these things!

I shall discuss many of the issues raised in this dialogue in the coming chapters. The essence of the dispute is whether one is simply to accept the explosive appearance of the universe as a bald, unexplained fact—something belonging to the "that's-that" category—or to seek some more satisfying explanation. Until recently it seemed as if any such explanation would have to involve a supernatural agency who transcended the laws of physics. But then a new advance was made in our understanding of the very early universe that has transformed the entire debate, and recast this age-old puzzle in a totally different light.

Creation without Creation

Since the demise of the steady-state theory, scientists have seemed to be faced with a stark choice concerning the origin of the universe. One could either believe that the universe is infinitely old, with all the attendant physical paradoxes, or else assume an abrupt origin of time (and space), the explanation for which lies beyond the scope of science. What was overlooked was a third possibility: that time can be bounded in the past and yet not come into existence abruptly at a singularity.

Before getting into the details of this, let me make the general point that the essence of the origin problem is that the big bang seems to be an event without a physical cause. This is usually regarded as contradicting the laws of physics. There is, however, a tiny loophole. This loophole is called quantum mechanics. As explained in chapter 1, the application of quantum mechanics is normally restricted to atoms, molecules, and subatomic particles. Quantum effects are usually negligible for macroscopic objects. Recall that at the heart of quantum physics lies Heisenberg's uncertainty principle, which states that all measurable quantities (e.g., position, momentum, energy) are subject to unpredictable fluctuations in their values. This unpredictability implies that the microworld is indeterministic: to use Einstein's picturesque phraseology, God plays dice with the universe. Therefore, quantum events are not determined absolutely by preceding causes. Although the probability of a given event (e.g., the radioactive decay of an atomic nucleus) is fixed by the theory, the actual outcome of a particular quantum process is unknown and, even in principle, unknowable.

By weakening the link between cause and effect, quantum mechanics provides a subtle way for us to circumvent the origin-of-the-universe problem. If a way can be found to permit the universe to come into existence from nothing as the result of a quantum fluctuation, then no laws of physics would be violated. In other words, viewed through the eyes of a quantum physicist, the spontaneous appearance of a universe is not such a surprise, because physical objects are spontaneously appearing all the time—without well-defined causes—in the quantum microworld. The quantum physicist need no more appeal to a supernatural act to bring the universe into being than to explain why a radioactive nucleus decayed when it did.

All of this depends, of course, on the validity of quantum mechanics when applied to the universe as a whole. This is not clear-cut. Quite apart from the astonishing extrapolation involved in applying a theory of subatomic particles to the entire cosmos, there are deep questions of principle concerning the meaning to be attached to certain mathematical objects in the theory. But many distinguished physicists have argued that the theory can be made to work satisfactorily in this situation, and thus was the subject of "quantum cosmology" born.

The justification for quantum cosmology is that, if the big bang is taken seriously, there would have been a time when the universe was compressed to minute dimensions. Under these circumstances quantum processes must have been important. In particular the fluctuations described by Heisenberg's uncertainty principle must have had a profound effect on the structure and evolution of the nascent cosmos. A simple calculation tells us when that epoch was. Quantum effects were important when the density of matter was a staggering 10^{94} gm cm^{-3}. This state of affairs existed before 10^{-43} seconds, when the universe was a mere 10^{-33} cm across. These numbers are referred to as the Planck density, time, and distance respectively, after Max Planck, the originator of the quantum theory.

The ability of quantum fluctuations to "fuzz out" the physical world on an ultramicroscopic scale leads to a fascinating prediction concerning the nature of space-time. Physicists can observe quantum fluctuations in the laboratory down to distances of about 10^{-16} cm and over times of about 10^{-26} seconds. These fluctuations affect such things as the positions and momenta of particles, and they take place within an apparently fixed space-time background. On the much smaller Planck scale, however, the fluctuations would also affect space-time itself.

To understand how, it is first necessary to appreciate the close linkage between space and time. The theory of relativity requires that we view three-dimensional space and one-dimensional time as parts of a unified four-dimensional space-time. In spite of the unification, space remains physically distinct from time. We have no difficulty in distinguishing them in daily life. This distinction can become blurred, however, by quantum fluctuations. At the Planck scale the separate identities of space and time can be smeared out. Precisely how depends on the details of the theory, which can be used to compute the relative probabilities of various space-time structures.

It may happen, as a result of these quantum effects, that the most probable structure for space-time under some circumstances is actually four-dimensional space. It has been argued by James Hartle and Stephen Hawking that precisely those circumstances prevailed in the very early universe. That is, if we imagine going backward in time toward the big bang, then, when we reach about one Planck time after what we thought was the initial singularity, something peculiar starts to happen. Time begins to "turn into" space. Rather than having to deal with the origin of space-time, therefore, we now have to contend with four-dimensional space, and the question arises as to the shape of that space—i.e., its geometry. In fact, the theory permits an infinite variety of shapes. Which one pertained in the actual universe is related to the problem of choosing the right initial conditions, a subject that will be discussed in more detail shortly. Hartle and Hawking make a particular choice, which they claim is natural on grounds of mathematical elegance.

It is possible to give a helpful pictorial representation of their idea. The reader is cautioned, however, not to take the pictures too literally. The starting point is to represent space-time by a diagram with time drawn vertically and space horizontally (see figure 4). Future is toward

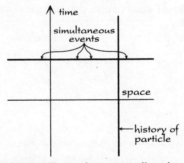

FIGURE 4. Space-time diagram. Time is drawn vertically and space horizontally. Only one dimension of space is shown. A horizontal section in the diagram represents all space at one instant of time. A vertical line represents a fixed point in space (e.g., the position of a stationary particle) throughout time.

the top of the diagram, past toward the bottom. Because it is impossible to represent four dimensions properly on the page of a book, I have eliminated all but one dimension of space, which is nevertheless adequate to make the essential points. A horizontal slice through the diagram represents all space at one instant of time, and a vertical line represents the history of a point in space at successive times. It is helpful to imagine having this diagram drawn on a sheet of paper on which certain operations can be performed. (The reader may find it instructive actually to carry these out.)

If space and time were infinite, we would, strictly speaking, need an infinite sheet of paper for our diagram to represent space-time properly. However, if time is bounded in the past, then the diagram must have a boundary somewhere along the bottom: one can imagine cutting a horizontal edge somewhere. It may also have a future boundary, demanding a similar edge along the top. (I have denoted these by the wiggly horizontal lines in figure 5.) In that case we would have an infinite strip representing all of infinite space at successive moments from the beginning of the universe (at the bottom edge) to the end (at the top edge).

At this stage one can entertain the possibility that space is not, after all, infinite. Einstein was the first to point out that space might be finite yet unbounded, and the idea remains a serious and testable cosmological hypothesis. Such a possibility is readily accommodated in our picture by rolling the sheet around to make a cylinder (figure 6). Space at each instant is now represented by a circle of finite circumference. (The two-dimensional analogue is the surface of a sphere; in three

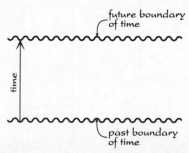

FIGURE 5. It may be that time is bounded by singularities in the past and/or future. This is represented on a space-time diagram by truncating the diagram at the bottom and top respectively. The wiggly lines denote the singularities.

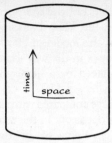

FIGURE 6. It may be that space is finite yet without boundary. This is represented by rolling the space-time diagram into a cylinder. A horizontal section, representing space at one instant, is then a circle.

dimensions it is a so-called hypersphere, which is hard to imagine but mathematically perfectly well defined and understood.)

A further refinement is to introduce the expansion of the universe, which may be represented by letting the size of the universe change with time. As we are concerned here with the origin of the universe, I shall ignore the top edge of the diagram, and show only that portion near the bottom. The cylinder has now become cone-shaped; a few circles are drawn to represent the expanding volume of space (figure 7). The hypothesis that the universe originated in a singularity of infinite compression is depicted here by allowing the cone to taper to a single point at the base. The singular apex of the cone represents the abrupt appearance of both space and time in a big bang.

The essential claim of quantum cosmology is that the Heisenberg

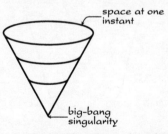

FIGURE 7. Expanding universe. The effect of cosmological expansion can be represented on our space-time diagram by making the cylinder of figure 6 into a cone. The apex of the cone corresponds to the big-bang singularity. Horizontal sections through the cone are now circles of successively greater diameter, denoting space growing larger.

uncertainty principle smears out the sharpness of the apex, replacing it by something smoother. Just what that something is depends on the theoretical model, but in the model of Hartle and Hawking a rough guide is to round off the apex in the way shown in figure 8, where the point of the cone is replaced by a hemisphere. The radius of this hemisphere is the Planck length (10^{-33} cm), very small by human standards, but infinitely large compared with a point singularity. Above this hemisphere the cone opens out in the usual way, representing the standard nonquantum development of the expanding universe. Here—in the upper portion, above the join to the hemisphere—time runs vertically up the cone as usual, and is physically quite distinct from space, which runs horizontally around the cone. Below the join, however, the situation is dramatically different. The time dimension starts to curve around into the space direction (i.e., the horizontal). Near the base of the hemisphere one has a two-dimensional, roughly horizontal curved surface. This represents a two-dimensional space rather than one space and one time dimension. Notice that the transition from time to space is gradual; it must not be thought to occur abruptly at the join. Expressing it another way, one might say that time emerges gradually from space as the hemisphere curves gradually into the cone. Note also that in this scheme time is still bounded from below—it does not stretch on back into the infinite past—yet there is no actual "first moment" of time, no abrupt beginning at a singular origin. The big-bang singularity has, in fact, been abolished.

P

FIGURE 8. Creation without creation. In this version of the origin of the universe the apex of the cone of figure 7 is rounded off. There is no abrupt beginning: time fades gradually away toward the base of the diagram. The event P looks like the first moment, but this is only an artifact of the way the diagram is drawn. There is no well-defined beginning, although time is still finite in the past.

One might still be tempted to think of the base of the hemisphere—the "South Pole"—as the "origin" of the universe, but, as Hawking emphasizes, this is mistaken. A portion of a spherical surface is characterized by the fact that, geometrically, all points on it are equivalent. That is, no point is singled out as privileged in any way. The base of the hemisphere looks special to us because of the way that we have chosen to represent the curved sheet. If the cone is tipped over a bit, some other point becomes the "base" of the structure. Hawking points out that the situation is somewhat analogous to the way we represent the spherical surface of the Earth geometrically. Lines of latitude converge on the North and South poles, but the Earth's surface at these places is the same as anywhere else. We could equally well have picked Mecca or Hong Kong as the focus for these circles. (The actual choice has been dictated by the axis of rotation of the Earth, a feature which is irrelevant to the present discussion.) There is no suggestion that the surface of the Earth comes to an abrupt stop at the poles. There is, to be sure, a singularity in the coordinate system of latitude and longitude there, but not a physical singularity in the geometry.

To make this point clearer, imagine making a little hole at the "South Pole" of the hemisphere in figure 8, then opening the sheet out around the hole (suppose it is elastic) to make a cylinder, then unwrapping the cylinder and spreading it out to form a flat sheet. We would end up with a picture just like figure 5. The point is that what we formerly took to be a singular origin to time (the bottom edge) is really just the coordinate singularity at the South Pole, infinitely stretched out. Exactly the same thing happens with maps of the Earth in Mercator's projection. The South Pole, which is really just a perfectly ordinary point on the Earth's surface, is represented by a horizontal boundary line, as though the Earth's surface has an edge there. But the edge is purely an artifact of the way we have chosen to represent the spherical geometry by a particular coordinate system. We are free to redraw a map of the Earth using a different coordinate system, with some other point chosen as the focus for circles of latitude, in which case the South Pole would appear on the map as it really is—a perfectly normal point.

The upshot of all this is that, according to Hartle and Hawking, there *is* no origin of the universe. Nevertheless, that does not mean that the universe is infinitely old. Time is limited in the past, but has no boundary as such. Thus centuries of philosophical anguish over the paradoxes of infinite versus finite time are neatly resolved. Hartle and Hawking ingeniously manage to pass between the horns of that particular dilemma. As Hawking expresses it: "The boundary condition of the universe is that it has no boundary."[5]

The implications of the Hartle-Hawking universe for theology are profound, as Hawking himself remarks: "So long as the universe had a beginning, we could suppose it had a creator. But if the universe is completely self-contained, having no boundary or edge, it would have neither beginning nor end: it would simply be. What place, then, for a creator?"[6] The argument is therefore that, because the universe does not have a singular origin in time, there is no need to appeal to a supernatural act of creation at the beginning. The British physicist Chris Isham, himself an expert on quantum cosmology, has made a study of the theological implications of the Hartle-Hawking theory. "There is no doubt that, psychologically speaking, the existence of this initial singular point is prone to generate the idea of a Creator who sets the whole show rolling," he writes.[7] But these new cosmological ideas remove the need, he believes, to invoke a God-of-the-gaps as the cause of the big bang: "The new theories would appear to plug *this* gap rather neatly."

Although Hawking's proposal is for a universe without a definite origin in time, it is also true to say in this theory that the universe has not always existed. Is it therefore correct to say that the universe has "created itself"? The way I would rather express it is that the universe of space-time and matter is internally consistent and self-contained. Its existence does not require anything outside of it; specifically, no prime mover is needed. So does this mean that the existence of the universe can be "explained" scientifically without the need for God? Can we regard the universe as forming a closed system, containing the reason for its existence entirely within itself? The answer depends on the meaning to be attached to the word "explanation." Given the laws of physics, the universe can, so to speak, take care of itself, including its own creation. But where do these laws come from? Must we, in turn, find an explanation for *them*? This is a topic I will take up in the next chapter.

Can these recent scientific developments square with the Christian doctrine of creation *ex nihilo*? As I have repeatedly emphasized, the idea of God bringing the universe into existence from nothing cannot be regarded as a temporal act, because it involves the creation of time. In the modern Christian viewpoint, creation *ex nihilo* means sustaining the universe in existence at all times. In modern scientific cosmology, one should no longer think of space-time as "coming into existence" anyway. Rather, one says that space-time (or the universe) simply *is*. "This scheme does not have an initial event with a special status," remarks the philosopher Wim Drees. "Hence, all moments have a similar relation to the Creator. Either they are all 'always there,' as a brute fact, or they are all equally created. It is a nice feature of this quantum cosmology that that part of the content of creation *ex nihilo* which was supposed to be the most decoupled from science, namely the 'sustaining,' can be seen as the more natural part in the context of the theory."[8] The image of God conjured up by this theory, however, is rather far removed from the twentieth-century Christian God. Drees perceives a close resemblance to the pantheistic picture of God adopted by the seventeenth-century philosopher Spinoza, where the physical universe itself takes on aspects of God's existence, such as being "eternal" and "necessary."

One can, of course, still ask: Why does the universe exist? Should the (timeless) existence of space-time be regarded as an (atemporal) form of "creation"? In this sense creation "from nothing" would not refer to any temporal transition from nothing to something, but would merely serve as a reminder that there could have been nothing rather than something. Most scientists (though perhaps not all—see page 124) would agree that the existence of a mathematical scheme for a universe is not the same thing as the actual existence of that universe. The scheme still has to be implemented. Thus there remains what Drees calls "ontological contingency." The Hartle-Hawking theory fits this more abstract sense of "creation" rather well, because it is a quantum theory. The essence of quantum physics, as I have remarked, is uncertainty: prediction in a quantum theory is prediction of *probabilities* rather than certainties. The Hartle-Hawking mathematical formalism supplies the *probabilities* that a particular universe, with a particular arrangement of matter, exists at each moment. In predicting that there is a nonzero probability for some particular universe, one is saying that there is a definite chance that it will be actualized. Thus creation *ex nihilo* is here given the concrete interpretation of "actualization of possibilities."

Mother and Child Universes

Before leaving the problem of the origin of the universe, I should say something about a recent cosmological theory in which the question of origin enters in a radically different way. In my book *God and the New Physics* I floated the idea that what we call the universe might have started out as an outgrowth of some larger system, which then detached itself to become an independent entity. The basic idea is illustrated in figure 9. Here space is represented as a two-dimensional sheet. In accordance with the general theory of relativity, we can imagine this sheet as curved. In particular, one can conceive of a localized bump forming on the sheet, and rising into a protuberance connected to the main sheet by a thin throat. It may then happen that the throat becomes progressively narrower, until it pinches off completely. The protuberance has then turned into a completely disconnected "bubble." The "mother" sheet has given rise to a "child."

Amazingly, there is good reason to expect something like this to be going on in the real universe. The random fluctuations associated with quantum physics imply that, on an ultramicroscopic scale, all manner of bumps, wormholes, and bridges should be forming and collapsing throughout space-time. The Soviet physicist Andrei Linde has the idea that our universe started out this way, as a little bubble of space-time, which then "inflated" at a fantastic rate to produce a big bang. Others have developed similar models. The "mother" universe which spawned ours is also continuously inflating at a fantastic rate, and spewing out baby universes for all it is worth. If this state of affairs is correct, it implies that "our" universe is only part of an infinite assemblage of universes, although it is self-contained now. The assemblage as a whole has no beginning or end. There are problems in any case in using words like "beginning" and "end," because there is no suprauniversal time in which this spawning process takes place, although each bubble has its own internal time.

An interesting question is whether our universe is also capable of being mother, and producing child universes. Might it be possible for some mad scientist to create his or her very own universe in the laboratory? This question has been investigated by Alan Guth, the originator of the inflationary theory. It turns out that, if a large amount of energy is concentrated, a space-time protuberance might indeed form. At first sight this seems to raise the alarming prospect that a new big

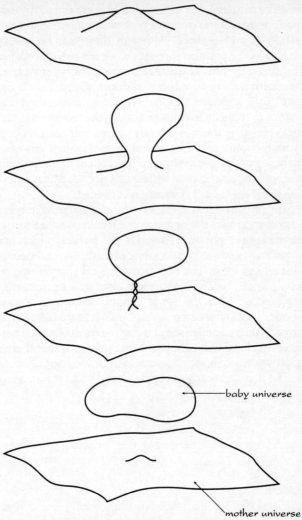

baby universe

mother universe

FIGURE 9. Hatching a baby universe. The mother universe is represented by a two-dimensional sheet. Curvature in the sheet arises from gravitational effects. If gravity is intense enough, the curvature can produce a protuberance that forms a mini-universe connected by an umbilical cord or throat known as a "wormhole." From the mother universe the throat can appear as a black hole. Eventually the hole evaporates, severing the cord and dispatching the baby universe onto an independent existence.

bang would be triggered, but in fact what happens is that the formation of the protuberance looks from our region of space-time to be just like the creation of a black hole. Although there may be an explosive inflation within the protuberance-space, we see only a black hole that steadily shrinks. Eventually the hole evaporates away completely, and at that moment our universe becomes disconnected with its child.

In spite of the appeal of this theory, it remains highly speculative. I shall return to it briefly in chapter 8. Both the mother-and-child and the Hartle-Hawking theories adroitly circumvent the problems associated with a cosmic origin by appealing to quantum processes. The lesson to be learned is that quantum physics opens the door to universes of a finite age, the existence of which does not demand a well-defined prior cause. No special act of creation is needed.

All of the physical ideas discussed in this chapter have been based on the assumption that the universe as a whole complies with certain well-defined laws of physics. These laws of physics, which underpin physical reality, are woven into a fabric of mathematics, itself founded on the bedrock of logic. The path from physical phenomena, through the laws of physics, to mathematics and ultimately logic, opens up the beguiling prospect that the world can be understood through the application of logical reasoning alone. Could it be that much, if not all, of the physical universe might be as it is as the result of logical necessity? Some scientists have indeed claimed that this is so, that there is only one logically consistent set of laws and only one logically consistent universe. To investigate this sweeping claim, we must inquire into the nature of the laws of physics.

3

What Are the Laws of Nature?

IN CHAPTER 2 I argued that, given the laws of physics, the universe can create itself. Or, stated more correctly, the existence of a universe without an external first cause need no longer be regarded as conflicting with the laws of physics. This conclusion is based, in particular, on the application to cosmology of quantum physics. Given the laws, the existence of the universe is not itself miraculous. This makes it seem as if the laws of physics act as the "ground of being" of the universe. Certainly, as far as most scientists are concerned, the bedrock of reality can be traced back to these laws. They are the eternal truths upon which the universe is built.

The concept of law is so well established in science that until recently few scientists stopped to think about the nature and origin of these laws; they were happy to simply accept them as "given." Now that physicists and cosmologists have made rapid progress toward finding what they regard as the "ultimate" laws of the universe, many old questions have resurfaced. Why do the laws have the form they do? Might they have been otherwise? Where do these laws come from? Do they exist independently of the physical universe?

The Origin of Law

The concept of a law of nature wasn't invented by any particular philosopher or scientist. Although the idea was crystallized only in the modern scientific era, its origins go back to the dawn of history, and are intimately bound up with religion. Our distant ancestors must have had a rudimentary notion of cause and effect. The purpose of making tools, for example, has always been to facilitate the manipulation of the environment. Hitting a nut with a stone causes it to break open, and a carefully thrown spear can be aimed with confidence that it will follow a particular trajectory. But although certain regularities of behavior were apparent to these early people, the vast majority of natural phenomena remained mysterious and unpredictable, and gods were invented to explain them. Thus there were the rain god, the sun god, tree gods and river gods, and so on. The natural world was under the control of a plethora of unseen powerful beings.

There is always a danger in judging earlier cultures on our own terms, with all our tacit assumptions and prejudices. In the age of science we regard it as perfectly natural to seek mechanistic explanations of things: the bow string propels the arrow, gravity pulls the stone toward the ground. A given cause, usually in the form of a force, produces a later effect. But early cultures did not generally regard the world in this way. Some perceived nature as a battleground of conflicting forces. Gods or spirits, each with a distinctive personality, would clash or compromise. Other cultures, especially in the East, believed that the physical world was a holistic tapestry of interdependent influences.

In almost all early cosmological theories, the world was likened not to a machine but to a living organism. Physical objects were endowed with purposes, much as animals seem to behave purposively. A vestige of this thinking survives today, when people talk about water "seeking out" its lowest level, or refer to the end of a compass needle as "north-seeking." The idea of a physical system seeking out, being directed, or being drawn toward some final goal is known as "teleology." The Greek philosopher Aristotle, whose animistic picture of the universe I mentioned briefly in chapter 1, distinguished between four sorts of causes: Material Cause, Formal Cause, Efficient Cause, and Final Cause. These categories are often illustrated by the example of a house. What causes a house to come into being? First there is the Material Cause, which is here identified with the bricks and other stuff from which the house is made. Then there is the Formal Cause, which is the form or shape that the material is arranged in. Third comes the Efficient Cause, the means whereby the material is fashioned into its form (in this case the builder). Lastly there is the

Final Cause, the purpose of the thing. In the case of a house this purpose might involve a pre-existing blueprint toward which the builder works.

Even armed with a fairly elaborate notion of causation, Aristotle did not properly formulate what we would today understand as laws of nature. He discussed the motion of material bodies, but his so-called laws of motion were really just descriptions of how Final Causes were supposed to operate. So, for example, a stone would fall because the "natural place" of ponderous objects was the Earth, tenuous gases would rise because their natural place was in the ethereal realm above the sky, and so on.

Much of this early thinking was based on the assumption that the properties of physical things were intrinsic qualities belonging to those things. The great diversity of forms and substances found in the physical world thus reflected the limitless variety of intrinsic properties. Set against this way of looking at the world were the monotheistic religions. The Jews conceived of God as the Lawgiver. This God, being independent of and separate from his creation, imposed laws upon the physical universe from without. Nature was supposed to be subject to laws by divine decree. One could still assign causes to phenomena, but the connection between cause and effect was now constrained by the laws. John Barrow has studied the historical origins of the concept of physical laws. He contrasts the Greek pantheon with the One monarchical God of Judaism: "When we look at the relatively sophisticated society of Greek gods, we do not find the notion of an all-powerful cosmic lawgiver very evident. Events are decided by negotiation, deception, or argument rather than by omnipotent decree. Creation proceeds by committee rather than fiat."[1]

The view that laws are imposed upon, rather than inherent in, nature was eventually adopted by Christianity and Islam too, though not without a struggle. Barrow relates how Saint Thomas Aquinas "viewed the innate Aristotelian tendencies as aspects of the natural world which were providentially employed by God. However, in this cooperative enterprise their basic character was inviolate. According to this view, God's relationship with Nature is that of a partner rather than that of a sovereign."[2] But such Aristotelian ideas were condemned by the Bishop of Paris in 1277, to be replaced in later doctrine by the notion of God the Lawmaker, so well encapsulated in Kempthorn's hymn of 1796:

Praise the Lord! for He hath spoken
Worlds his mighty voice obeyed
Laws, which never shall be broken
For their guidance He hath made.

It is fascinating to trace the cultural and religious influences at work in the formulation of the modern concept of laws of nature. Medieval Europe, subject on the one hand to the Christian doctrine of God's law manifested in nature, and on the other hand to a strongly enforced concept of civil law, provided a fertile milieu for the scientific idea of laws of nature to emerge. Thus we find that the early astronomers such as Tycho Brahe and Johannes Kepler, in deducing the laws of planetary motion, believed that in studying the orderly processes of nature they were uncovering God's rational design. This position was further articulated by the French philosopher-scientist René Descartes, and adopted by Isaac Newton, whose laws of motion and gravitation gave birth to the age of science.

Newton himself believed strongly in a Designer who worked through fixed mathematical laws. For Newton and his contemporaries the universe was a vast and magnificent machine constructed by God. Opinions differed, however, as to the nature of the Cosmic Mathematician and Engineer. Did he merely construct the machine, wind it up, and leave it to look after itself? Or did he actively supervise its running on a day-to-day basis? Newton believed that the universe was saved from gravitational disintegration only by a perpetual miracle. Such divine intervention is a classic example of the God-of-the-gaps. It is an argument fraught with danger, and leaves hostage to fortune that a future advance in science may satisfactorily plug the gap. And, indeed, the gravitational stability of the universe is today well understood. Even in Newton's day, his perpetual-miracle assumption was derided by his Continental rivals. Thus Leibniz taunted:

Mr. Newton and his followers have also an extremely odd opinion of the work of God. According to them God has to wind up his watch from time to time. Otherwise it would cease to go. He lacked sufficient foresight to make it a perpetual motion. . . . According to my view, the same force and vigour goes on existing in the world always.[3]

For Descartes and Leibniz, God was the fountainhead and guarantor of the total rationality that pervades the cosmos. It is this rationality that opens the door to the understanding of nature by the application of human reason, itself a gift from God. In Renaissance Europe, the justification for what we today call the scientific approach to inquiry was the belief in a rational God whose created order could be discerned from a careful study of nature. And, Newton notwithstanding, part of this belief came to be that God's laws were immutable. "The scientific culture that arose in Western Europe," writes Barrow, "of which we are the inheritors, was dominated by adherence to the absolute invariance of laws of Nature, which thereby underwrote the meaningfulness of the scientific enterprise and assured its success."[4]

For the modern scientist, it is sufficient only that nature simply have the observed regularities we still call laws. The question of their origin does not usually arise. Yet it is interesting to ponder whether science would have flourished in medieval and Renaissance Europe were it not for Western theology. China, for example, had a complex and highly developed culture at that time, which produced some technological innovations that were in advance of Europe's. The Japanese scholar Kowa Seki, who lived at the time of Newton, is credited with the independent invention of the differential calculus and a procedure for computing pi, but he chose to keep these formulations secret. In his study of early Chinese thought, Joseph Needham writes: "There was no confidence that the code of Nature's laws could ever be unveiled and read, because there was no assurance that a divine being, even more rational than ourselves, had ever formulated such a code capable of being read."[5] Barrow argues that, in the absence of "the concept of a divine being who acted to legislate what went on in the natural world, whose decrees formed inviolate 'laws' of Nature, and who underwrote scientific enterprise," Chinese science was condemned to a "curious stillbirth."[6]

Although there is undoubtedly some truth in the claim that differences in scientific progress between East and West can be traced to theological differences, other factors are also responsible. The greater part of Western science has been founded on the method of reductionism, whereby the properties of a complicated system are understood by studying the behavior of its component parts. To give a simple example, there is probably nobody who understands all the systems of a Boeing 747 airliner, but every part of it is understood by somebody.

We are happy to say that the airliner's behavior as a whole is understood, because we believe that an airliner is just the sum of its parts.

Our ability to dissect natural systems in this way has been crucial to the progress of science. The word "analysis," often used synonymously with "science," expresses this assumption that we can take things apart and study the bits in isolation in order to understand the whole. Even a system as complex as the human body, it is claimed by some, can be understood by knowing the behavior of the individual genes, or the rules governing the molecules that make up our cells. If we could not understand limited parts of the universe without understanding the whole, science would be a hopeless enterprise. Yet this analyzable quality of physical systems is not so universal as once thought. In recent years scientists have come to recognize more and more systems that must be understood holistically or not at all. These systems are described mathematically by equations known as "nonlinear." (More details can be found in my books *The Cosmic Blueprint* and *The Matter Myth.*) It may simply be an accident of history that the first scientists were preoccupied with linear physical systems, like the solar system, which are especially amenable to analytical techniques and a reductionist approach.

The popularity of "holistic science" in recent years has prompted a string of books, most notably Fritjof Capra's *The Tao of Physics*, that stress the similarities between ancient Eastern philosophy, with its emphasis on the holistic interconnectedness of physical things, and modern nonlinear physics. Can we conclude that Oriental philosophy and theology were, after all, superior to their Western counterparts? Surely not. We now appreciate that scientific progress requires both reductionistic and holistic approaches. It is not a question of one being right and the other wrong, as some people like to assert, but the need for two complementary ways of studying physical phenomena. What is interesting is that reductionism works at all. Why is it that the world is structured in such a way that we can know something without knowing everything? This is a topic that I shall pursue in chapter 6.

The Cosmic Code

The rise of science and the Age of Reason brought with it the idea of a hidden order in nature, which was mathematical in form and could be uncovered by ingenious investigation. Whereas, in primitive considerations of cause and effect, direct connections are immediately apparent to the senses, the laws of nature discovered by science are altogether more subtle. Anyone can see, for example, that apples fall, but Newton's inverse-square law of gravitation demands special and systematic measurement before it is manifested. More important, it demands some sort of abstract theoretical framework, evidently of a mathematical nature, as a context for those measurements. The raw data gathered by our senses are not directly intelligible as they stand. To link them, to weave them into a framework of understanding, requires an intermediary step, a step we call theory.

The fact that such theory is subtle and mathematical can be suggestively expressed by saying that the laws of nature are in code. The job of the scientist is to "crack" the cosmic code, and thereby reveal the secrets of the universe. Heinz Pagels, in his book *The Cosmic Code*, expresses it thus:

> Although the idea that the universe has an order that is governed by natural laws that are not immediately apparent to the senses is very ancient, it is only in the last three hundred years that we have discovered a method for uncovering the hidden order—the scientific-experimental method. So powerful is this method that virtually everything scientists know about the natural world comes from it. What they find is that the architecture of the universe is indeed built according to invisible universal rules, what I call the cosmic code—the building code of the Demiurge.[7]

As explained in chapter 1, Plato envisaged a benevolent craftsman—a Demiurge—who built the universe using mathematical principles based on symmetric geometrical forms. This abstract realm of Platonic Forms was connected with the everyday world of sense experiences by a subtle entity Plato called the World-Soul. The philosopher Walter Mayerstein likens Plato's World-Soul to the modern concept of mathematical theory, being the thing that connects our sense experiences with the principles on which the universe is built, and provides us with what we call understanding.[8] In the modern era, Einstein also insisted that our direct observations of events in the world are not generally intelligible, but must be related to a layer of underlying theory. In a letter to M. Solovine dated 7 May 1952, Einstein wrote of "the always problematical connection between the world of ideas and that which can be experienced." Einstein stresses that there is "no logical path" between the theoretical concepts and our observations. One is brought into concordance with the other by an "extra-logical (intuitive)" procedure.[9]

Using a computer metaphor, we might say that the laws of nature encode a message. We are the receivers of that message, communicated to us through the channel we call scientific theory. For Plato, and many others after him, the emitter of this message is the Demiurge, the cosmic Builder. As we shall see in the forthcoming chapters, all information about the world can in principle be represented in the form of binary arithmetic (ones and zeros), this being the form most convenient for computer processing. "The universe," claims Mayerstein, "can be simulated as an enormous string of 0's and 1's; the purpose of the scientific endeavour then being nothing else than the attempt at decoding and un-scrambling this sequence with the objective of trying to understand, to make sense of this 'message.' " What can be said about the nature of the "message"? "Quite obviously, if the message is coded, this presupposes the existence of some pattern or structure in the arrangement of 0's and 1's in the string; a thoroughly random or chaotic string must be considered to be undecodable."[10] So the fact that there is cosmos rather than chaos boils down to the patterned properties of this string of digits. In chapter 6 I shall inquire further into the exact nature of these properties.

The Status of the Laws Today

Many people, including some scientists, would like to believe that the cosmic code contains a real message for us from an Encoder. They maintain that the very existence of the code is evidence for the existence of an Encoder, and that the content of the message tells us something about him. Others, such as Pagels, find no evidence for an Encoder at all: "One of the odd features of the cosmic code is that, as far as we can tell, the Demiurge has written himself out of the code—an alien message without evidence of an alien." So the laws of nature have become a message without a Sender. Pagels is not unduly perturbed by this. "Whether God is the message, wrote the message, or whether it wrote itself is unimportant to our lives. We can safely drop the idea of a Demiurge, for there is no scientific evidence for a Creator of the natural world, no evidence for a will or purpose in nature that goes beyond the known laws of nature."[11]

As long as the laws of nature were rooted in God, their existence was no more remarkable than that of matter, which God also created. But if the divine underpinning of the laws is removed, their existence becomes a profound mystery. Where do they come from? Who "sent the message"? Who devised the code? Are the laws simply *there*—free-floating, so to speak—or should we abandon the very notion of laws of nature as an unnecessary hangover from a religious past?

To get a handle on these deep issues, let us first take a look at what a scientist actually means by a law. Everybody agrees that the workings of nature exhibit striking regularities. The orbits of the planets, for example, are described by simple geometrical shapes, and their motions display distinct mathematical rhythms. Patterns and rhythms are also found within atoms and their constituents. Even everyday structures such as bridges and machines generally behave in an ordered and predictable manner. On the basis of such experiences, scientists use inductive reasoning to argue that these regularities are lawlike. As explained in chapter 1, inductive reasoning has no absolute security. Just because the sun has risen every day of your life, there is no guarantee that it will therefore rise tomorrow. The belief that it will—that there are indeed dependable regularities of nature—is an act of faith, but one which is indispensable to the progress of science.

It is important to understand that the regularities of nature are real. Sometimes it is argued that laws of nature, which are attempts to capture these regularities systematically, are imposed on the world by our minds in order to make sense of it. It is certainly true that the human mind does have a tendency to spot patterns, and even to imagine them where none exist. Our ancestors saw animals and gods amid the stars, and invented the constellations. And we have all looked for faces in clouds and rocks and flames. Nevertheless, I believe any suggestion that the laws of nature are similar projections of the human mind is absurd. The existence of regularities in nature is an objective mathematical fact. On the other hand, the statements called laws that are found in textbooks clearly *are* human inventions, but inventions designed to reflect, albeit imperfectly, actually existing properties of nature. Without this assumption that the regularities are real, science is reduced to a meaningless charade.

Another reason why I don't think the laws of nature are simply made up by us is that they help us uncover new things about the world, sometimes things we never suspected. The mark of a powerful law is that it goes beyond a faithful description of the original phenomenon it was invoked to explain, and links up with other phenomena too. Newton's law of gravity, for example, gives an accurate account of planetary motion, but it also explains the ocean tides, the shape of the Earth, the motion of spacecraft, and much else. Maxwell's electromagnetic theory went far beyond a description of electricity and magnetism, by explaining the nature of light waves and predicting the existence of radio waves. The truly basic laws of nature thus establish deep connections between different physical processes. The history of science shows that, once a new law is accepted, its consequences are rapidly worked out, and the law is tested in many novel contexts, often leading to the discovery of new, unexpected, and important phenomena. This leads me to believe that in conducting science we are uncovering real regularities and linkages, that we are reading these regularities out of nature, not writing them into nature.

Even if we don't know what the laws of nature are, or where they have come from, we can still list their properties. Curiously, the laws have been invested with many of the qualities that were formally attributed to the God from which they were once supposed to have come.

First and foremost, the laws are universal. A law that only works sometimes, or in one place but not another, is no good. The laws are taken to apply unfailingly everywhere in the universe and at all epochs of cosmic history. No exceptions are permitted. In this sense they are also perfect.

Second, the laws are absolute. They do not depend on anything else. In particular they do not depend on who is observing nature, or on the actual state of the world. The physical states are affected by the laws, but not vice versa. Indeed, a key element in the scientific world view is the separation of the laws governing a physical system from the states of that system. When a scientist talks about the "state" of a system, she or he means the actual physical condition that the system is in at some moment. To describe a state, you have to give the values of all the physical quantities that characterize that system. The state of a gas, for example, can be specified by giving its temperature, pressure, chemical composition, and so on, if you are interested only in its gross features. A complete specification of the state of the gas would mean giving details of the positions and motions of all the constituent molecules. The state is not something fixed and God-given; it will generally change with time. By contrast, the laws, which provide correlations between states at subsequent moments, do not change with time.

So we arrive at a third and most important property of the laws of nature: they are eternal. The timeless, eternal character of the laws is reflected in the mathematical structures employed to model the physical world. In classical mechanics, for example, the dynamical laws are embodied in a mathematical object called the "Hamiltonian" which acts in something called "phase space." These are technical mathematical constructions; their definition is not important. What matters is that both the Hamiltonian and the phase space are fixed. On the other hand, the state of the system is represented by a point in phase space, and this point moves about with time, representing the changes of state that occur as the system evolves. The essential fact is that the Hamiltonian and the phase space themselves are independent of the motion of the representative point.

Fourth, the laws are omnipotent. By this I mean that nothing escapes them: they are all-powerful. They are also, in a loose sense, omniscient, for, if we go along with the metaphor of the laws "commanding" physical systems, then the systems do not have to "inform" the laws of their states in order for the laws to "legislate the right instructions" for that state.

This much is generally agreed. A schism appears, however, when we consider the status of the laws. Are they to be regarded as discoveries about reality, or merely as the clever inventions of scientists? Is Newton's inverse-square law of gravity a discovery about the real world that happens to have been made by Newton, or is it an invention of Newton's made in an attempt to describe observed regularities? Put differently, did Newton uncover something objectively real about the world, or did he merely invent a mathematical model of a part of the world that just happens to be rather useful in describing it?

The language that is used to discuss the operation of Newton's laws reflects a strong prejudice for the former position. Physicists talk about planets "obeying" Newton's laws, as though a planet were inherently a rebellious entity that would run amok if it were not "subject" to the laws. This gives the impression that the laws are somehow "out there," lying in wait, ready to supervene in the motions of planets whenever and wherever they occur. Falling into the habit of this description, it is easy to attribute an independent status to the laws. If they are considered to have such status, then the laws are said to be transcendent, because they transcend the actual physical world itself. But is this really justified?

How can the separate, transcendent existence of laws be established? If laws manifest themselves only through physical systems—in the way that physical systems behave—we can never get "behind" the stuff of the cosmos to the laws as such. The laws are *in* the behavior of physical things. We observe the things, not the laws. But if we can never get a handle on the laws except through their manifestation in physical phenomena, what right have we got to attribute to them an independent existence?

A helpful analogy here is with the concepts of hardware and software in computing. The laws of physics correspond to software, the physical states to hardware. (Granted, this stretches the use of the word "hard" quite a bit, as included in the definition of the physical universe are nebulous quantum fields and even space-time itself.) The foregoing issue can then be stated thus: Is there an independently existing "cosmic software"—a computer program for a universe—encapsulating all the necessary laws? Can this software exist without the hardware?

I have already indicated my belief that the laws of nature are real, objective truths about the universe, and that we discover them rather than invent them. But all known fundamental laws are found to be mathematical in form. Why this should be so is an important and subtle topic that requires an investigation of the nature of mathematics. This I shall take up in the coming chapters.

What Does It Mean for Something to "Exist"?

If physical reality is somehow built on the laws of physics, then these laws must have an independent existence in some sense. What form of existence can we attribute to something so abstract and nebulous as a law of nature?

Let me start with something concrete—like concrete, for instance. We know it exists, because (in the famous words of Dr. Johnson) we can kick it. We can also see it and possibly smell it: concrete directly affects our senses. But there is more to the existence of a lump of concrete than feeling, sight, and smell. We also make the assumption that the existence of the concrete is something that is independent of our senses. It is really "out there," and will go on existing even when we don't touch, see, or smell it. This is, of course, a hypothesis, but a reasonable one. What actually happens is that, on repeated inspection, we receive similar sense data. The correlation between the sense data received on successive occasions enables us to recognize the lump of concrete and identify it. It is then simpler to make our model of reality on the basis that the concrete has independent existence, than to suppose it vanishes when we look away and obligingly reappears each time we look back.

All this seems uncontentious. But not all things said to exist are as concrete as concrete. How about atoms, for example? They are too small to see or touch, or to sense directly in any way. Our knowledge of them comes indirectly, via intermediary equipment, the data from which must be processed and interpreted. Quantum mechanics makes things worse. It is not possible, for example, to attribute a definite position as well as a definite motion to an atom at the same time. Atoms and subatomic particles inhabit a shadowy world of half-existence.

Then there are still more abstract entities such as fields. The gravitational field of a body certainly exists, but you cannot kick it, let alone see or smell it. Quantum fields are still more nebulous, consisting of quivering patterns of invisible energy.

But less-than-concrete existence is not the preserve of physics. Even in daily life we use concepts such as citizenship or bankruptcy, which, although they can't be touched or seen, are nevertheless very real. Another example is information. The fact that information as such cannot be directly sensed does not diminish the real importance in our lives of "information technology," in which information is stored and processed. Similar remarks apply to the concept of software, and software engineering, in the science of computing. Of course, we might be able to see or touch the medium of information storage, such as a computer disc or microchip, but we cannot directly perceive the information on these as such.

Then there is the whole realm of subjective phenomena, such as dream images. Dream objects undeniably enjoy a kind of existence (at least for the dreamer), but of an altogether less substantial nature than lumps of concrete. Similarly for thoughts, emotions, memories, and sensations: they cannot be dismissed as nonexistent, although the nature of their existence is different from that of the "objective" world.

Like computer software, the mind or soul might depend for its *manifestation* on something concrete—in this case the brain—but that does not make it concrete.

There is also a category of things that are broadly described as cultural—music, for example, or literature. The existence of the symphonies of Beethoven or the works of Dickens cannot simply be equated with the existence of the manuscripts on which they are written. Nor can religion or politics be identified merely with the people who practice it. All these things "exist" in a less-than-concrete, but nevertheless important, sense.

Finally, there is the realm of mathematics and logic, of central concern to science. What is the nature of their existence? When we say that there exists a certain theorem about, say, prime numbers, we do not mean that the theorem can be kicked, like the lump of concrete. Yet mathematics undeniably has existence of some sort, albeit abstract.

The question that confronts us is whether the laws of physics enjoy a transcendent existence. Many physicists believe that this is so. They talk about the "discovery" of the laws of physics as though these are already "out there" somewhere. Of course, it is conceded that what we today call the laws of physics are only a tentative approximation to a unique set of "true" laws, but the belief is that as science progresses so these approximations get better and better, with the expectation that one day we will have the "correct" set of laws. When this happens, theoretical physics will be complete. It was the expectation that just such a culmination lies in the not-too-distant future that prompted Stephen Hawking to entitle his inaugural lecture to the Lucasian Chair in Cambridge "Is the End in Sight for Theoretical Physics?"

Not all theoretical physicists are so comfortable with the idea of transcendent laws, however. James Hartle, observing that "scientists, like mathematicians, proceed as though the truths of their subjects had an independent existence . . . as though there were a single set of rules by which the universe is run with an actuality apart from this world they govern," argues that the history of science is replete with examples of how what were once regarded as indispensable fundamental truths turned out to be both dispensable and special.[12] That the Earth was the center of the universe went unquestioned for centuries until we found that the universe only seemed that way on account of our location on its surface. That lines and angles in three-dimensional space obey the laws of Euclidean geometry was also assumed as a fundamental and indispensable truth, but turned out to be due only to the fact that we live in a region of space and time in which gravity is relatively weak, so that the curvature of space went unnoticed for a long time. How many other features of the world, wonders Hartle, might similarly be due to our particular perspective on the world, and not the result of a deep transcendent truth? The separation of nature into "the world" and "the laws" might be one such dispensable feature.

According to this point of view, there is no unique set of laws toward which science converges. Our theories, and the laws contained therein, cannot be separated, says Hartle, from the circumstances in which we find ourselves. These circumstances include our culture and evolutionary history, and the specific data we have collected about the world. An alien civilization with a different evolutionary history, culture, and science might construct very different laws. Hartle points out that many different laws can be fitted to a given set of data, and that we can never be sure that we have attained the correct set.

In the Beginning

It is important to realize that laws do not by themselves completely describe the world. Indeed, the whole purpose of our formulating laws is to connect different physical events. It is a simple law, for example, that a ball thrown in the air will follow a parabolic path. However, there are many different parabolas. Some are tall and thin, others low and shallow. The particular parabola followed by a particular ball will depend on the speed and angle of projection. These are referred to as "initial conditions." The parabola law plus the initial conditions determine the path of the ball uniquely.

The laws, then, are statements about classes of phenomena. Initial conditions are statements about particular systems. In conducting his or her science, the experimental physicist will often choose, or contrive, certain initial conditions. For example, in his famous experiment on falling bodies, Galileo released unequal masses simultaneously, in order to demonstrate that they strike the ground at the same moment. By contrast, the scientist cannot choose the laws; they are "God-given." This fact imbues the laws with a much higher status than the initial conditions. The latter are regarded as an incidental and malleable detail, whereas the former are fundamental, eternal, and absolute.

In the natural world, outside the realm of the experimenter's control, the initial conditions are provided for us by nature. The hailstone that strikes the ground was not dropped by Galileo in some predetermined manner, but was produced by physical processes in the upper atmosphere. Similarly, when a comet enters the solar system from without along a particular path, that path depends on the physical processes of the comet's origin. In other words, the initial conditions pertaining to a system of interest can be traced to the wider environment. One can then ask about the initial conditions of that wider environment. Why did the hailstone form at that particular point in the atmosphere? Why did the clouds form there rather than somewhere else? And so on.

It is easy to see that the web of causal interconnections spreads outward very rapidly until it encompasses the entire cosmos. What then? The question of the cosmic initial conditions leads us back to the big bang and the origin of the physical universe. Here the rules of the game change dramatically. Whereas for a particular physical system the initial conditions are just an incidental feature that can always be explained by appealing to the wider environment at an earlier moment, when it comes to the cosmic initial conditions there *is* no wider environment, and no earlier moment. The cosmic initial conditions are "given," just like the laws of physics.

Most scientists regard the cosmic initial conditions as lying outside the scope of science altogether. Like the laws, they must simply be accepted as a brute fact. Those of a religious frame of mind appeal to God to explain them. Atheists tend to regard them as random or arbitrary. It is the job of the scientist to explain the world as far as possible without appeal to special initial conditions. If some feature of the world can be accounted for only by supposing that the universe began in a certain way, then no real explanation has been given at all. One is merely saying that the world is the way it is because it was the way it was. The temptation has therefore been to construct theories of the universe that do not depend very sensitively on the initial conditions.

A clue to how this can be done is provided by thermodynamics. If I am given a cup of hot water, I know it will be cold the next day. On the other hand, if I am given a cup of cold water, I can't say whether or not it was hot the day before, or the day before that, or how hot, or whether it was ever hot at all. One might say that the details of the thermal history of the water, including its initial conditions, are erased by the thermodynamic processes that bring it into thermal equilibrium with its environment. Cosmologists have argued that similar processes could have erased the details of the cosmic initial conditions. It would then be impossible to infer, except in the broadest terms, how the universe began simply from a knowledge of what it is like today.

Let me give an example. The universe is expanding today at the same rate in every direction. Does this mean that the big bang was isotropic? Not necessarily. It could have been the case that the universe started out expanding in a chaotic way, with different rates in different directions, and that this disorder was smoothed out by physical processes. For instance, frictional effects could act to brake the motion in the directions of rapid expansion. Alternatively, according to the fashionable inflationary-universe scenario* discussed briefly in chapter 2, the early universe underwent a phase of accelerating expansion in which all initial irregularities were stretched out of existence. The end result was a universe with a high degree of spatial uniformity and a smooth pattern of expansion.

Many scientists are attracted to the idea that the state of the universe we observe today is relatively insensitive to the way it started out in the big bang. No doubt this is partly due to a reaction against religious theories of special creation, but it is also because the idea removes the need to worry about the state of the universe in its very early stages, when the physical conditions were likely to have been extreme. On the other hand, it is clear that initial conditions can't be completely ignored. We can imagine a universe of the same age as ours but of very different form, and then envisage it being evolved backward in time in accordance with the laws of physics to a big-bang origin. Some initial state would be discovered which would then give rise to that different universe.

Whatever initial conditions gave rise to our universe, one can always ask: Why those? Given the infinite variety of ways in which the universe could have started out, why did it start out in the way it did? Is there something special, perhaps, about those particular initial conditions? It is tempting to suppose that the initial conditions were not arbitrary, but conformed to some deep principle. After all, it is usually accepted that the laws of physics are not arbitrary, but can be

---

* For a detailed account of this theory, see my book *Superforce*.

encapsulated in neat mathematical relationships. Might not there exist a neat mathematical "law of initial conditions" too?

Such a proposal has been advanced by a number of theorists. Roger Penrose, for example, has argued that, if the initial conditions were chosen at random, the resulting universe is overwhelmingly likely to be highly irregular, containing monster black holes rather than relatively smoothly distributed matter. A universe as smooth as ours requires some extraordinarily delicate fine-tuning at the outset, so that all regions of the universe expand in a carefully orchestrated manner. Using the metaphor of the Creator with a limitless "shopping list" of possible initial conditions, Penrose points out that the Creator would need to peruse the list very thoroughly before finding a candidate that would lead to a universe like ours. Sticking in a pin at random would be a strategy almost certain to fail. "Without wishing to denigrate the Creator's abilities in this respect," remarks Penrose, "I would insist that it is one of the duties of science to search for physical laws which explain, or at least describe in some coherent way, the nature of the phenomenal accuracy that we so often observe in the workings of the natural world. . . . So we need a law of physics to explain the spe-cialness of the initial state."[13] The law proposed by Penrose is that the initial state of the universe was constrained to possess a specific type of smoothness right from the outset, without any need for inflation or other smoothing processes. The mathematical details need not con-cern us.

Another proposal has been discussed by Hartle and Hawking in the context of their quantum-cosmological theory. In chapter 2 I men-tioned that there is no particular "first moment" in this theory, no creation event. The problem of the cosmic initial conditions is there-fore abolished by abolishing the initial event altogether. However, to achieve this end, the quantum state of the universe must be severely restricted, not just at the beginning, but at all times. Hartle and Hawking give a definite mathematical formulation of such a restriction, which in effect plays the role of a "law of initial conditions."

It is important to realize that a law of initial conditions can't be proved right or wrong, or derived from existing laws of physics. The value of any such law rests, as with all scientific proposals, in its ability to predict observable consequences. True, theorists may be attracted to a particular proposal on grounds of mathematical elegance and "nat-uralness," but such philosophical arguments are hard to justify. The

Hartle-Hawking proposal, for example, is well adapted to the formalism of quantum gravity, and seems very plausible and natural within that context. But had our science developed differently, the Hartle-Hawking law might have appeared highly arbitrary or contrived.

Unfortunately, pursuing the observational consequences of the Hartle-Hawking theory isn't easy. The authors claim that it predicts an inflationary phase for the universe, which accords with the latest cosmological fashion, and it might one day have something to say about the large-scale structure of the universe—the way in which galaxies tend to cluster together, for example. But there seems little hope of ever selecting a unique law on observational grounds. Indeed, Hartle has argued (see page 168) that no such unique law exists. In any case, a given proposal to select a quantum state of the entire universe will not have very much to say about the fine level of detail, such as the existence of a particular planet, still less a particular person. The very quantum nature of the theory ensures (because of Heisenberg's uncertainty principle) that such details remain indeterminate.

The separation into laws and initial conditions that has characterized all past attempts to analyze dynamical systems might owe more to the history of science than to any deep property of the natural world. The textbooks tell us that in a typical experiment the experimenter creates a particular physical state and then observes what happens—i.e., how the state evolves. The success of the scientific method rests on the reproducibility of the results. If the experiment is repeated, the same laws of physics apply, but the initial conditions are under the control of the experimenter. There is thus a clear functional separation between laws and initial conditions. When it comes to cosmology, however, the situation is different. There is only one universe, so the notion of repeated experimentation is inapplicable. Moreover, we have no more control over the cosmic initial conditions than we do over the laws of physics. The sharp distinction between the laws of physics and the initial conditions therefore break down. "Is it not possible," conjectures Hartle, "that there are some more general principles in a more general framework which determine both the initial conditions and dynamics?"[14]

I believe that these proposals about laws of initial conditions strongly support the Platonic idea that the laws are "out there," transcending the physical universe. It is sometimes argued that the laws of physics came into being with the universe. If that was so, then those laws cannot explain the origin of the universe, because the laws would not exist until the universe existed. This is most forcefully obvious when it comes to a law of initial conditions, because such a law purports to explain precisely how the universe came to exist in the form that it does. In the Hartle-Hawking scheme there is no actual moment of creation at which their law applies. Nevertheless, it is still proposed as an explanation for why the universe has the form it does. If the laws are not transcendent, one is obliged to accept as a brute fact that the universe is simply *there*, as a package, with the various features described by the laws built in. But with transcendent laws one has the beginnings of an explanation for why the universe is as it is.

The idea of transcendent laws of physics is the modern counterpart of Plato's realm of perfect Forms which acted as blueprints for the construction of the fleeting shadow-world of our perceptions. In practice, the laws of physics are framed as mathematical relationships, so in our search for the bedrock of reality we must now turn to the nature of mathematics, and to the ancient problem of whether mathematics exists in an independent Platonic realm.

4

Mathematics and Reality

NO SUBJECT BETTER ILLUSTRATES the divide between the two cultures—arts and sciences—than mathematics. To the outsider, mathematics is a strange, abstract world of horrendous technicality, full of weird symbols and complicated procedures, an impenetrable language and a black art. To the scientist, mathematics is the guarantor of precision and objectivity. It is also, astonishingly, the language of nature itself. No one who is closed off from mathematics can ever grasp the full significance of the natural order that is woven so deeply into the fabric of physical reality.

Because of its indispensable role in science, many scientists—especially physicists—invest the ultimate reality of the physical world in mathematics. A colleague of mine once remarked that in his opinion the world was *nothing but* bits and pieces of mathematics. To the ordinary person, whose picture of reality is tied closely to the perception of physical objects, and whose view of mathematics is that of an esoteric recreation, this must seem astounding. Yet the contention that mathematics is a key that enables the initiate to unlock cosmic secrets is as old as the subject itself.

Magic Numbers

Mention ancient Greece and most people think of geometry. Today children learn Pythagoras' theorem and the other elements of Euclidean geometry as a training exercise for mathematical and logical thought. But to the Greek philosophers their geometry represented much more than an intellectual exercise. The concepts of number and shape fascinated them so deeply that they constructed an entire theory of the universe upon it. In Pythagoras' words: "Number is the measure of all things."

Pythagoras himself lived in the sixth century B.C. and founded a school of philosophers known as the Pythagoreans. They were convinced that the cosmic order was based upon numerical relationships, and they imbued certain numbers and forms with mystical significance. They had special reverence, for example, for the "perfect" numbers such as 6 and 28, which are the sum of their divisors (e.g., $6 = 1 + 2 + 3$). Greatest respect was reserved for the number 10, the so-called divine tetraktus, being the sum of the first four whole numbers. By arranging dots into various shapes, they constructed triangular numbers (such as 3, 6, and 10), square numbers (4, 9, 16, etc.), and so on. The square number 4 was made the symbol for justice and reciprocity, a meaning that retains a faint echo in the expressions "a square deal" and "being all square." The triangular representation of 10 was regarded as a sacred symbol, and sworn upon during initiation ceremonies.

The Pythagoreans' belief in the power of numerology was bolstered by Pythagoras' discovery of the role of number in music. He found that the lengths of strings that produced harmonically related tones bore simple numerical relationships to each other. The octave, for example, corresponded to the ratio 2:1. Our word "rational" ("ratio-nal") derives from the great heuristic significance that the Pythagoreans gave to numbers obtained as the ratio of whole numbers, such as ¾ or ⅔. Indeed, mathematicians still refer to such numbers as rational. It was therefore deeply unsettling to the Greeks when they discovered that the square root of 2 *cannot* be expressed as the ratio of whole numbers. What does this mean? Imagine a square with each side measuring one meter. Then, according to Pythagoras' own theorem, the length of the diagonal in meters is the square root of 2. This length is roughly ⅞ meters; a better approximation is ⁷⁰⁷/₅₀₀ meters. But in fact there is *no* exact fraction that can express it, however big the numerator and denominator are allowed to be. Numbers of this sort are still called "irrational."

The Pythagoreans applied their numerology to astronomy. They devised a system of 9 concentric spherical shells to convey the known heavenly bodies as they turned, and invented a mythical "counter-Earth" to make up the tetraktus number 10. This connection between musical and heavenly harmony was epitomized by the assertion that the astronomical spheres gave forth music as they turned—the music of the spheres. Pythagorean ideas were endorsed by Plato, who in his *Timaeus* developed further a musical and numerical model of the cosmos. He went on to apply numerology to the Greek elements—earth, air, fire, and water—and to explore the cosmic significance of various regular geometrical forms.

The Pythagorean and Platonic schemes seem primitive and eccentric to us today, although I do from time to time receive manuscripts in the mail containing attempts to explain the properties of atomic nuclei, or subnuclear particles, on the basis of early Greek numerology. Evidently it retains a certain mystical appeal. The main value of these numerological and geometrical systems is not, however, their plausibility, but the fact that they treat the physical world as a manifestation of concordant mathematical relationships. This essential idea survived right up to the scientific era. Kepler, for example, described God as a geometer, and in his analysis of the solar system was profoundly influenced by what he perceived to be the mystical significance of the numbers involved. And modern mathematical physics, though divested of mystical overtones, nevertheless retains the ancient Greek assumption that the universe is rationally ordered according to mathematical principles.

Numerological schemes have been developed by many other cultures and have penetrated both science and art. In the ancient Near East the number 1—the Unity—was often identified with God the Prime Mover. The Assyrians and Babylonians assigned deified numbers to astronomical objects: Venus, for example, was identified with the number 15, and the moon with 30. The Hebrews placed special significance on 40, which recurs often in the bible. The devil is associated with 666, a number that retains some potency today if, as a journalist once reported, President Ronald Reagan altered his Californian address to avoid it. In fact, the bible has numerology woven deep into its fabric, both in its content and in the organization of the text. Some later religious sects, such as the Gnostics and Cabalists, constructed elaborate and esoteric numerological lore around the bible. Nor was the Church immune from such theorizing. Augustine in particular encouraged the numerological study of the bible as part of a

Christian education, and this practice persisted until the late Middle Ages. In our own time many cultures continue to ascribe supernatural powers to certain numbers or geometrical shapes, and special counting routines form an important part of ritual and magic in many parts of the world. Even in our skeptical Western society many people hang on to the notion of lucky or unlucky numbers, such as 7 or 13.

These magical connotations obscure the very practical origins of arithmetic and geometry. The construction of formal geometrical theorems in ancient Greece followed the development of the ruler and the compass, and various line-of-sight surveying techniques, which were used for architectural and construction purposes. From these simple technological beginnings a great system of thought was built. The power of number and geometry proved so compelling that it became the basis for a complete world view, with God himself cast in the role of the Great Geometer—the image captured so vividly in William Blake's famous etching *The Ancient of Days*, showing God stooping from the heavens to measure the universe with dividers.

History suggests that each age appeals to its most impressive technology as a metaphor for the cosmos, or even God. So it was that by the seventeenth century the universe was no longer regarded in terms of musical and geometrical harmony presided over by a cosmic Geometer, but in an altogether different way. An outstanding technological challenge at that time was the provision of accurate navigational devices, especially to assist in the European colonization of America. The determination of latitude presents no problems for navigators, because it can be directly measured by the altitude of, say, the polestar above the horizon. Longitude, however, is a different matter, because as the Earth rotates, so the heavenly bodies move across the sky. A position measurement must be combined with a time measurement. For east-west navigation, needed for transatlantic crossings, accurate clocks were essential. Thus, with the powerful motivation of political and commercial reward, much effort was devoted to designing precision timepieces for use at sea.

This focus on accurate timekeeping found its theoretical counterpart in the work of Galileo and Newton. Galileo used time as a parameter to establish his law of falling bodies. He is also credited with the discovery that the period of a pendulum is independent of the amplitude of its swing, a fact he is said to have established in church by timing the sway of a lantern against his pulse. Newton recognized the central role that time plays in physics, stating in his *Principia* that "absolute, true and mathematical time, of itself, and from its own nature, flows equably without relation to anything external."[1] Thus time, like distance, was acknowledged as a feature of the physical universe to be measured, in principle to arbitrary precision.

The further contemplation of the role of the flux of time in physics led Newton to develop his mathematical theory of "fluxions," today known as the calculus. The central feature of this formalism is the notion of continuous change. Newton made this the basis of his theory of mechanics, in which the laws of motion of material bodies were set down. The most striking and successful application of Newton's mechanics was to the motion of the planets in the solar system. Thus the music of the spheres was replaced by the image of the clockwork universe. This image achieved its most developed form with the work of Pierre Laplace in the late eighteenth century, who envisaged every atom in the universe as a component in an unfailingly precise cosmic clockwork mechanism. God the Geometer became God the Watchmaker.

Mechanizing Mathematics

Our own age has likewise enjoyed a technological revolution that is already coloring our entire world view. I refer to the rise of the computer, which has caused a profound shift in the way that both scientists and nonscientists alike think about the world. As in previous ages, so today there are suggestions that the latest technology be used as a metaphor for the operation of the cosmos itself. Thus some scientists have proposed we regard nature as basically a *computational process.* The music of the spheres and the clockwork universe have been displaced in favor of the "cosmic computer," with the entire universe being regarded as a gigantic information-processing system. According to this view, the laws of nature can be identified with the program of the computer, and the unfolding events of the world become the cosmic output. The initial conditions at the origin of the universe are the input data.

Historians now recognize that the modern concept of the computer can be traced back to the pioneering work of the eccentric English inventor Charles Babbage. Born near London in 1791, Babbage was the son of a wealthy banker whose family came from Totnes in Devonshire. Even as a child the young Babbage was interested in mechanical devices. He taught himself mathematics from whatever books came to hand, and went up to Cambridge University as a student in 1810 with his individual approach to the subject already established, and full of plans to challenge the orthodoxy of British mathematical teaching. Together with his lifelong friend John Herschel, son of the famous astronomer William Herschel (who discovered the planet Uranus in 1781), Babbage founded the Analytical Society. The Analyticals were greatly enamored of the power of French science and engineering, and saw the introduction of Continental-style mathematics in Cambridge as the first step in a technological and manufacturing revolution. The Society came into collision with the Cambridge political establishment, who regarded Babbage and company as militant radicals.

After leaving Cambridge, Babbage married and settled down in London, living on private means. He continued to admire French scientific and mathematical thought, possibly as a result of his acquaintance with the Bonaparte family, and he formed many scientific contacts on the Continent. At this stage he became interested in experimenting with calculating machines, and succeeded in securing government financing to construct what he called a Difference Engine, a type of adding machine. The purpose was to produce mathematical, astronomical, and navigational tables free from human error and with less labor. Babbage exhibited a small-scale working model of the Difference Engine, but the British government suspended funding in 1833, and the full-scale machine was never completed. This must have been one of the earliest examples of a government failing to recognize the need for long-term research support. (And I'm bound to say that, in Britain at least, little seems to have changed since the 1830s.) In the event, a Difference Engine was produced in Sweden, based on Babbage's design, and subsequently purchased by the British government.

Undaunted by this lack of support, Babbage conceived of a far more powerful computing device, a general-purpose computer which he termed the Analytical Engine, and which is now recognized as the forerunner of the modern computer in its essential organization and architecture. He spent much of his personal fortune attempting to construct several different versions of this Engine, but none was ever fully completed.

Babbage was a forceful, argumentative, and controversial figure, and many of his contemporaries dismissed him as a crank. Nevertheless, he is credited with the invention of, among other things, the speedometer, the ophthalmoscope, the cowcatcher for trains, the overhead cash trolley for shops, and coded flashing for lighthouses. His interests encompassed politics, economics, philosophy, and astronomy. Babbage's insight into the nature of computational processes led him to speculate that the universe could be regarded as a type of computer, with the laws of nature playing the role of program—a remarkably prescient speculation, as we shall see.

In spite of his eccentricity, Babbage's talents were duly recognized when he was elected to the Lucasian Chair of Mathematics in Cambridge, a post once held by Newton. As a historical footnote, two of Babbage's sons emigrated to Adelaide in South Australia, bringing pieces of the Engines with them. Meanwhile, back in London, a full-scale reconstruction of a Difference Engine has been made at the Science Museum. It has been assembled according to Babbage's original design to prove that it really can compute as intended. And in 1991, the bicentenary of Babbage's birth (shared, incidentally, with Faraday's birth and Mozart's death) was commemorated by Her Majesty's government with a special issue of postage stamps.

Following Babbage's death in 1871, his work was largely forgotten, and it was not until the 1930s, and the imagination of another unusual Englishman, Alan Turing, that the story moves on. Turing and the American mathematician John von Neumann are both credited with laying the logical foundations for the modern computer. Central to their work was the notion of a "universal computer," a machine capable of executing any computable mathematical function. To explain the significance of universal computation, one has to go back to 1900, to a famous address made by the mathematician David Hilbert in which he set out what he regarded as the twenty-three most important outstanding mathematical problems to be tackled. One of these concerned the question of whether a general procedure could be found for proving mathematical theorems.

Hilbert was aware that the nineteenth century had witnessed some deeply disturbing mathematical developments, some of which seemed to threaten the entire consistency of mathematics. These included problems associated with the concept of infinity, and various logical paradoxes of self-reference that I shall discuss shortly. In response to these doubts Hilbert challenged mathematicians to find a systematic procedure for deciding, in a finite number of steps, whether a given mathematical statement was true or false. Nobody at the time seemed to doubt that such a procedure should exist, although to construct it was quite another matter. Nevertheless, one could envisage the possibility of a person or a committee testing each mathematical conjecture by blindly following a prescribed sequence of operations through to the bitter end. Indeed, people would be irrelevant, because the procedure could be mechanized, and the machine made to follow the sequence of operations automatically, eventually halting and printing out the result—"true" or "false," as the case may be.

Viewed this way, mathematics becomes an entirely formal discipline, even a game, concerned only with manipulating symbols according to certain specified rules and establishing tautological relationships. It need have no relevance to the physical world. Let us see how this is so. When we carry out an arithmetic procedure such as $(5 \times 8) - 6 = 34$ we follow a simple set of rules to obtain the answer 34. To get the right answer, we do not need to understand the rules, or where they come from. In fact, we don't even need to understand what the symbols, such as 5 and $\times$, actually mean. As long as we correctly recognize the symbols and stick to the rules, we get the right answer. The fact that we can use a pocket calculator to do the job for us proves that the procedure can be implemented entirely blindly.

When children first learn arithmetic, they need to relate the symbols back to concrete objects in the real world, so they start by associating numbers with fingers, or beads. In later years, however, most children are happy to carry out mathematical operations entirely abstractly, even to the extent of using x and y in place of specific numbers. Those who go on to advanced mathematics learn about other types of numbers (e.g., complex) and operations (e.g., matrix multiplication), which obey strange rules that don't obviously correspond to anything familiar in the real world. Still, the student can readily learn how to manipulate the abstract symbols that denote these unfamiliar objects and operations, without worrying about what, if anything, they actually mean. So mathematics becomes more and more a question of the formal manipulation of symbols. It begins to seem as if mathematics is *nothing but* symbol manipulation, a point of view known as "formalism."

In spite of its superficial plausibility, the formalist interpretation of mathematics received a severe blow in 1931. In that year the Princeton mathematician and logician Kurt Gödel proved a sweeping theorem to the effect that mathematical statements existed for which *no* systematic procedure could determine whether they are either true or false. This was a no-go theorem with a vengeance, because it provided an irrefutable demonstration that something in mathematics is actually impossible, even in principle. The fact that there exist *undecidable propositions* in mathematics came as a great shock, because it seemed to undermine the entire logical foundations of the subject.

Gödel's theorem springs from a constellation of paradoxes that surround the subject of self-reference. Consider as a simple introduction to this tangled topic the disconcerting sentence: "This statement is a lie." If the statement is true, then it is false; and if it is false, then it is true. Such self-referential paradoxes are easily constructed and deeply intriguing; they have perplexed people for centuries. A medieval formulation of the same conundrum goes like this:

SOCRATES: "What Plato is about to say is false."
PLATO: "Socrates has just spoken truly."

(There are many versions: some references are given in the bibliography.) The great mathematician and philosopher Bertrand Russell demonstrated that the existence of such paradoxes strikes at the very heart of logic, and undermines any straightforward attempt to construct mathematics rigorously on a logical foundation. Gödel went on to adapt these difficulties of self-reference to the subject of mathematics in a brilliant and unusual manner. He considered the relationship between the *description* of mathematics and the mathematics itself. This is simple enough to state, but it actually required a long and very intricate argument. To get the flavor of what is involved, one can imagine listing mathematical propositions by labeling them 1, 2, 3. . . . Combining a sequence of propositions into a theorem then corresponds to combining the natural numbers that form their labels. In this way, logical operations *about* mathematics can be made to correspond to mathematical operations themselves. And this is the essence of the self-referential character of Gödel's proof. By identifying the subject with the object—mapping the description of the mathematics onto the mathematics—he uncovered a Russellian paradoxical loop that led directly to the inevitability of undecidable propositions. John Barrow has remarked wryly that, if a religion is defined to be a system of thought which requires belief in unprovable truths, then mathematics is the only religion that can prove it is a religion!

The key idea at the heart of Gödel's theorem can be explained with the help of a little story. In a faraway country a group of mathematicians who have never heard of Gödel become convinced that there does indeed exist a systematic procedure to determine infallibly the truth or falsity of every meaningful proposition, and they set out to demonstrate it. Their system can be operated by a person, or a group of people, or a machine, or a combination of any of these. Nobody was quite sure of what the mathematicians chose, because it was located in a large university building rather like a temple, and entry was forbidden to the general public. Anyway, the system was called Tom. To test Tom's abilities, all sorts of complicated logical and mathematical statements were presented to it, and, after due time for processing, back came the answers: true, true, false, true, false. . . . It was not long before Tom's fame spread throughout the land. Many people came to visit the laboratory, and exercised greater and greater ingenuity in formulating ever more difficult problems in an attempt to stump Tom. Nobody could. So confident grew the mathematicians of Tom's infallibility that they persuaded their king to offer a prize to anyone who could defeat Tom's incredible analytical powers. One day a traveler from another country came to the university with an envelope, and asked to challenge Tom for the prize. Inside the envelope was a piece of paper with a statement on it, intended for Tom. The statement, which we can give the name "S" ("S" for "statement" or "S" for "stump") simply read: "Tom cannot prove this statement to be true."

S was duly given to Tom. Scarcely had a few seconds elapsed before Tom began a sort of convulsion. After half a minute a technician came running from the building with the news that Tom had been shut down due to technical problems. What had happened? Suppose Tom were to arrive at the conclusion that S is true. This means that the statement "Tom cannot prove this statement to be true" will have been falsified, because Tom will have just done it. But if S is falsified, S cannot be true. Thus, if Tom answers "true" to S, Tom will have arrived at a false conclusion, contradicting its much-vaunted infallibility. Hence Tom cannot answer "true." We have therefore arrived at the conclusion that S is, in fact, true. But in arriving at this conclusion we have demonstrated that Tom cannot arrive at this conclusion. This means we know something to be true that Tom can't demonstrate to be true. This is the essence of Gödel's proof: that there will always exist certain true statements that cannot be proved to be true. The traveler, of course, knew this, and had no difficulty in constructing the statement S and claiming the prize.

It is important to realize, however, that the limitation exposed by Gödel's theorem concerns the axiomatic method of logical proof itself, and is not a property of the statements one is trying to prove (or disprove). One can always make the truth of a statement that is unprovable in a given axiom system *itself* an axiom in some extended system. But then there will be *other* statements unprovable in this enlarged system, and so on.

Gödel's theorem was a devastating setback for the formalist program, but the idea of a purely mechanical procedure for investigating mathematical statements was not completely abandoned. Perhaps undecidable propositions are just rare quirks that can be sifted out of logic and mathematics? If a way could be found to sort statements into decidable and undecidable, it might then still be feasible to determine for the former whether they were true or false. But could a systematic procedure be found for infallibly recognizing undecidable propositions and discarding them? The challenge of this task was taken up in the mid-1930s by Alonzo Church, a collaborator of von Neumann's at Princeton, and he soon demonstrated that even this more modest goal was unattainable, at least in a finite number of steps. That is to say, potentially true or false mathematical statements could be made, and a systematic procedure embarked upon to check their truthfulness or otherwise, but this procedure would never terminate: the result could never be known.

The Uncomputable

The problem was also addressed, quite independently and from a totally different slant, by Alan Turing, while he was still a young student in Cambridge. Mathematicians often speak of a "handle-turning" or "mechanical" procedure for solving mathematical problems. What fascinated Turing was whether an actual machine could be designed which would fulfill that function. Such a machine might then be capable of deciding the truth of mathematical statements automatically, without human involvement, by slavishly following a deterministic sequence of instructions. But what would be the structure of this machine? How would it work? Turing envisaged something like a typewriter, capable of marking symbols on a page, but having the additional quality of being able to read, or scan, other given symbols, and to erase them if necessary. He settled on the idea of a tape of indefinite length, divided into squares, with each square carrying a single symbol. The machine would move the tape one square at a time, read the symbol, and then either remain in the same state, or move to a new state, depending on what it read. In each case its response would be purely automatic, and determined by the construction of the machine. The machine would either leave the symbol alone, or erase it and type another, then move the tape by one square and continue.

In essence, the Turing machine is simply a device for transforming one string of symbols into another string according to a predetermined set of rules. These rules could, if necessary, be tabulated, and the behavior of the machine at each step read off from the table. There was no need actually to build a machine out of paper tape and metal or whatever to elucidate its capabilities. It is easy, for example, to work out a table that corresponds to an adding machine. But Turing was interested in more ambitious goals. Could his machine tackle Hilbert's program for the mechanization of mathematics?

As already remarked, solving mathematical problems by following a mechanical procedure is well drilled into schoolchildren. Converting a fraction into a decimal and obtaining a square root are favorites. Any finite set of manipulations leading to an answer—say, in the form of a number (not necessarily a whole number)—could clearly be handled by a Turing machine. But what about infinitely long procedures? The decimal expansion of pi, for example, is unending and seemingly random. Nevertheless, pi may be computed to any desired number of decimal places by following a simple finite rule. Turing called a number "computable" if, using a finite set of instructions, the number could be generated in this way to unlimited accuracy, even if the complete answer would be infinitely long.

Turing imagined a list of all computable numbers. The list would, of course, itself be infinitely long, and at first sight it might seem as if every conceivable number would be included somewhere in the list. However, this is not so. Turing was able to demonstrate that such a list could be used to discover the existence of other numbers which could not possibly be present anywhere in the list. As the list includes all computable numbers, it follows that these new numbers must be uncomputable. What does an uncomputable number mean? From its definition, it is a number that cannot be generated by a finitely defined mechanical procedure, even by executing an infinite number of steps. Turing had shown that a list of computable numbers could be used to generate uncomputable numbers.

Here is the gist of his argument. Imagine that instead of numbers we are dealing with names. Consider listing six-letter names: Sayers, Atkins, Piquet, Mather, Belamy, Panoff, say. Now carry out the following simple procedure. Take the first letter of the first name and advance it alphabetically by one place. This gives "T." Then do the same for the second letter of the second name, the third letter of the third name, and so on. The result is "Turing." We can be absolutely certain that the name Turing cannot have been present in the original list, because it must differ from each name in that list by at least one letter. Even if we hadn't seen the original list, we would know that Turing could not be on it. Returning to the case of computable numbers, Turing used a similar one-change-in-each-number argument to show the existence of uncomputable numbers. Of course, Turing's list contained an infinite number of infinitely long numbers, rather than six six-letter words, but the essence of the argument is the same.

The existence of uncomputable numbers already suggests that there must be undecidable mathematical propositions. Imagine the infinite list of computable numbers. Each number can be generated by a Turing machine. One machine could be constructed to calculate a square root, another a logarithm, and so on. As we just saw, this could never produce all numbers, even with an infinity of such machines, because of the existence of uncomputable numbers, which cannot be generated mechanically. Turing spotted that it was not in fact necessary to have an infinity of Turing machines to generate this list. Only one was necessary. He demonstrated that it was possible to construct a *universal* Turing machine, which is capable of simulating all other Turing machines. The reason why such a universal machine can exist is simple. Any machine can be specified by giving a systematic procedure for its construction: washing machines, sewing machines, adding machines, Turing machines. The fact that a Turing machine is itself a machine to carry out a procedure is the key point. Hence a universal Turing machine can be instructed to first read off the specification of any given Turing machine, then reconstruct its internal logic, and finally execute its function. Clearly, then, the possibility exists of a general-purpose machine capable of performing all mathematical tasks. One no longer needs an adding machine to add, a multiplying machine to multiply, and so on. A single machine could do it all. This was implicit in Charles Babbage's proposal for his Analytical Engine, but it took nearly a century, the genius of Alan Turing, and the demands of the Second World War for the concept of the modern computer finally to come of age.

It might seem astonishing that a machine which can simply read, write, erase, move, and stop is capable of exploring all conceivable mathematical procedures, irrespective of how abstract or complicated they are. Nevertheless, this claim, called the Church-Turing hypothesis, is believed by most mathematicians. It means that, whatever the mathematical problem concerned, if a Turing machine can't solve it, nobody can. The Church-Turing hypothesis carries the important implication that it doesn't really matter what detailed construction is used for a computer. So long as it has the same basic logical structure as a universal Turing machine, the results will be the same. In other words, computers can simulate each other. Today, a real electronic computer is likely to have screen editing facilities, a printer, a graph plotter, disc storage, and other sophisticated devices, but its basic structure is that of the universal Turing machine.

When Turing was carrying out his analysis in the mid-thirties, all these important practical applications of his ideas lay in the future. His immediate concern was Hilbert's program for the mechanization of mathematics. The problem of computable and uncomputable numbers bears directly on this. Consider the (infinite) list of computable numbers, each generated by a Turing machine. Imagine the universal Turing machine being assigned the task of compiling this list on its own by successively simulating all Turing machines. The first step is to read off the construction details of each machine. A question then immediately arises: can the universal Turing machine tell from those details, in advance of actually executing the computation, whether a number could in fact be computed, or whether the computation would get stuck somewhere? Getting stuck means being trapped in some computational loop, failing to print any digits. This is known as the "halting problem"—whether it is possible to tell in advance, by inspecting the details of a computational procedure, if that procedure will compute each digit of some number and then halt, or if it will get trapped in a loop and never halt.

Turing showed that the answer to the halting problem is a decisive no. He did this using a clever argument. Suppose, he asked, that the universal machine could solve the halting problem. What would then happen if the universal machine were to attempt to simulate itself? We are back to the problems of self-reference. The result, as might be expected, is a computational seizure. The machine goes into an endless loop, chasing itself nowhere. So Turing arrived at a bizarre contradiction: the machine that is supposed to check in advance whether a computational procedure will get stuck in a loop gets stuck in a loop itself! Turing had exposed a variant of Gödel's theorem about undecidable propositions. In this case the undecidability concerns undecidable propositions themselves: there is no systematic way of deciding whether a given proposition is decidable or undecidable. Here, then, was a clear counterexample to Hilbert's conjecture about the mechanization of mathematics: a theorem that cannot be proved or disproved by a systematic general procedure. The profound nature of Turing's result was graphically summarized by Douglas Hofstadter: "Undecidable propositions run through mathematics like threads of gristle that criss-cross a steak in such a dense way that they cannot be cut out without the entire steak's being destroyed."[2]

Why Does Arithmetic Work?

These results of Turing's are usually interpreted as telling us something about mathematics and logic, but they also tell us something about the real world. The concept of a Turing machine, after all, is based upon our intuitive understanding of what a machine is. And real machines do what they do only because the laws of physics permit them to. Recently the Oxford mathematical physicist David Deutsch has claimed that computability is actually an *empirical* property, which is to say that it depends on the way the world happens to be rather than on some necessary logical truth. "The reason why we find it possible," writes Deutsch, "to construct, say, electronic calculators, and indeed why we can perform mental arithmetic, cannot be found in mathematics and logic. *The reason is that the laws of physics 'happen to' permit the existence of physical models for the operation of arithmetic such as addition, subtraction and multiplication. If they did not, these familiar operations would be non-computable functions.*"[3]

Deutsch's conjecture is certainly arresting. Arithmetic operations such as counting seem so basic to the nature of things that it seems hard to conceive of a world in which they could not be performed. Why is this? I think the answer might have something to do with the history and nature of mathematics. Simple arithmetic began with very mundane practical matters, like keeping track of sheep and basic accountancy. But the elementary operations of addition, subtraction, and multiplication triggered an explosive growth in mathematical ideas that eventually became so complex that people lost sight of the humble practical origins of the subject. In other words, mathematics took on a life and existence of its own. Already by the time of Plato some philosophers asserted that mathematics possessed an existence of its own. And we are so used to doing simple arithmetic that it is easy to believe it *must* be doable. But in fact its doability depends fundamentally on the nature of the physical world. For example, would counting make sense to us if there did not exist discrete objects such as coins and sheep?

The mathematician R. W. Hamming refuses to take the doability of arithmetic for granted, finding it both strange and inexplicable. "I have tried, with little success," he writes, "to get some of my friends to understand my amazement that the abstraction of integers for counting is both possible and useful. Is it not remarkable that 6 sheep plus 7 sheep make 13 sheep; that 6 stones plus 7 stones make 13 stones? Is it not a miracle that the universe is so constructed that such a simple abstraction as number is possible?"[4]

The fact that the physical world reflects the computational properties of arithmetic has a profound implication. It means that, in a sense, the physical world *is* a computer, as Babbage conjectured. Or, more to the point, computers can not only simulate each other, they can also simulate the physical world. Of course, we are perfectly familiar with the way that computers are used to model physical systems; indeed, that is their great utility. But this capability hinges on a deep and subtle property of the world. There is evidently a crucial *concordance* between, on the one hand, the laws of physics and, on the other hand, the computability of the mathematical functions that describe *those same laws*. This is by no means a truism. The nature of the laws of physics permits certain mathematical operations—such as addition and multiplication—to be computable. We find that among these computable operations are some which describe (at least to some accuracy) the laws of physics. I have symbolized this self-consistent loop in figure 10.

Is this loopy self-consistency just a coincidence, or does such consistency have to be the case? Does it point to some deeper resonance between mathematics and reality? Imagine a world in which the laws of physics were very different, possibly so different that discrete objects did not exist. Some of the mathematical operations that are computable in our world would not be so in this world, and vice versa. The equivalent of Turing machines might exist in this other world, but their structure and operation would be so completely different that it would be impossible for them to perform, say, basic arithmetic, though it may be able to perform computations in that world which computers in our world could never accomplish (such as solving Fermat's last theorem?).

Some interesting additional questions now arise: Would the laws of physics in this hypothetical alternative world be expressible in terms of the computable operations of that world? Or might it be the case that such self-consistency is possible only in a restricted class of worlds? Maybe in our world alone? Moreover, can we be sure that all aspects of our world *are* expressible in terms of computable operations? Might there not be physical processes that could *not* be simulated by a Turing machine? These further intriguing questions, which probe the link between mathematics and physical reality, will be examined in the next chapter.

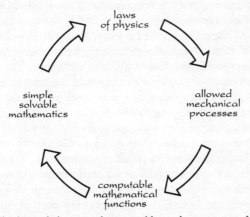

laws
of physics

simple
solvable
mathematics

allowed
mechanical
processes

computable
mathematical
functions

FIGURE 10. The laws of physics and computable mathematics may form a unique closed cycle of existence.

Russian Dolls and Artificial Life

The fact that universal computers can simulate each other has some important implications. On the practical level it means that, properly programmed and with enough memory space bolted on, a modest IBM PC can perfectly imitate, say, a powerful Cray computer as far as output (not speed) is concerned. Anything the Cray can do, so can the PC. In fact, a universal computer need be nowhere near as sophisticated as an IBM PC. It might consist of nothing more than a checkerboard and a supply of checkers! Such a system was first studied by the mathematicians Stanislaw Ulam and John von Neumann in the 1950s as an example of what is called "game theory."

Ulam and von Neumann were working at the Los Alamos National Laboratory, where the Manhattan atomic-bomb project was conducted. Ulam liked to play games on the computers, which were still a novelty at that time. One such game involved patterns that change shape according to certain rules. Imagine, for example, a checkerboard with checkers set out in some arrangement. One can then consider definite rules about how the pattern might be rearranged. Here is an example: Each square on the board has eight adjacent squares (including the diagonal neighbors). The state of any given square remains unchanged (i.e., with or without checker) if precisely two of the neighboring squares are occupied by checkers. If an occupied square has three occupied neighbors, it remains occupied. In all other cases the square becomes or remains empty. Some initial distribution of checkers is chosen, and the rule is applied to every square on the checkerboard. A slightly different pattern from the initial one is thereby obtained. The rule is then applied again, and further changes occur. The rule is then repeated and repeated, and the evolution of the pattern observed.

The particular rules given above were invented by John Conway in 1970, and he was instantly startled by the richness of the structures that resulted. Patterns appeared and disappeared, evolved, moved about, fragmented, merged. Conway was struck by the resemblance of these patterns to living forms, so he called the game "Life." Computer buffs the world over soon became addicted to it. They didn't need to use real checkerboards to follow the progress of the patterns. A less laborious procedure is to get a computer to display the patterns directly on a screen, with each pixel (dot of light) representing a checker. A delightfully readable account of the subject is given in the book *The*

Recursive Universe by William Poundstone.[5] An appendix gives a program for anyone wishing to play Life on his or her home computer. Owners of the Amstrad PCW 8256, the machine on which this book is being typed, may be interested to know that Life is already programmed into the machine and can be accessed with a few simple commands.

One can consider the space occupied by the dot patterns as a model universe, with Conway's rules substituting for the laws of physics, and time advancing in discrete steps. Everything that happens in the Life universe is strictly deterministic: the pattern at each step is completely determined by the pattern at the preceding step. The initial pattern thus fixes everything to come, *ad infinitum*. In this respect the Life universe resembles the Newtonian clockwork universe. Indeed, the mechanistic character of such games has earned them the name "cellular automata," the cells being the squares or pixels.

Among the infinite variety of Life forms are some that retain their identity as they move about. These include the so-called gliders, consisting of five dots, and various larger "spaceships." Collisions between these objects can produce all sorts of structures and debris, depending on the details. Gliders can be produced by a "glider gun," which emits them at regular intervals in a stream. Interestingly, glider guns can be made from thirteen-glider collisions, so that gliders beget gliders. Other common objects are "blocks," stationary squares of four dots that tend to destroy objects that collide with them. Then there are the more destructive "eaters," which break up and annihilate passing objects, and then repair the damage to themselves occasioned by the encounter. Conway and his colleagues have discovered Life patterns of immense richness and complexity, sometimes by chance, sometimes using great skill and insight. Some of the more interesting behavior demands careful choreography of large numbers of component objects, and shows up only after thousands of time steps. Very powerful computers are needed to explore the more advanced repertoire of Life activity.

The Life universe is obviously only a pale shadow of reality, the lifelike nature of its simpler inhabitants constituting merely a cartoon of real living things. Yet buried within its logical structure Life has the capacity to generate unlimited complexity, in principle as complex as genuine biological organisms. Indeed, von Neumann's original interest in cellular automata was closely connected with the mystery of living organisms. He was fascinated to know whether a machine could in principle be built that is capable of reproducing itself, and if so what its structure and organization might be. If such a von Neumann machine is possible, then we would be able to understand the principles that enable biological organisms to reproduce themselves.

The basis of von Neumann's analysis was the concept of a "universal constructor" analogous to a universal computer. This would be a machine that could be programmed to produce anything, much as a Turing machine can be programmed to execute any computable mathematical operation. Von Neumann considered what would happen if the universal constructor were programmed to make itself. Of course, to qualify for genuine self-reproduction a machine has to produce not only a copy of itself, but also a copy of the program of how to copy itself; otherwise the daughter machine will be "sterile." There is clearly a danger of an infinite regress here, but a clever evasion was spotted by von Neumann. The universal constructor must be augmented by a control mechanism. When the constructor has produced a copy of itself (plus a duplicate control mechanism, of course), the control mechanism switches off the program and treats it like just another bit of "hardware." The von Neumann machine duly makes a copy of the program, and inserts it into the new machine, which is then a faithful replica of the parent and ready to start running its own self-reproduction program.

Originally von Neumann had in mind a real "nuts-and-bolts" machine, but Ulam persuaded him to investigate the mechanistic possibilities of cellular automata, and to search for the existence of self-reproducing patterns. Von Neumann's "machine" might then be merely dots of light on a screen, or checkers on a checkerboard. No matter: it is the logical and organizational structure that is important, not the actual medium. After a lot of work, von Neumann and his colleagues were able to show that self-reproduction is indeed possible for systems that exceed a certain threshold of complexity. To do this required the investigation of a cellular automaton with rules considerably more complicated than those of the Life game. Rather than allowing each cell to be in only one of two states—empty or occupied— von Neumann's automaton permitted no fewer than twenty-nine alternative states. There was never any hope of actually constructing a self-reproducing automaton pattern—the universal constructor, control mechanism, and memory would have to occupy at least two hundred thousand cells—but the important point is that *in principle* a purely mechanistic system can reproduce itself. Shortly after this mathematical investigation was completed came the flowering of molecular biology, with the discovery of the double-helix structure of DNA, the unraveling of the genetic code, and the elucidation of the basic organization of molecular replication. It was soon realized that nature employs the same logical principles discovered by von Neumann. Indeed, biologists have identified the actual molecules within living cells that correspond to the components of a von Neumann machine.

Conway has been able to show that his Life game is also capable of permitting self-reproducing patterns. The relatively simple process of gliders making gliders does not qualify, because the all-important program for self-reproduction does not get copied. One needs something much more complicated for that. Conway first addressed a related question: can a Turing machine (i.e., a universal computer) be built in the Life universe? The basic operation of any universal computer consists of the logical operations AND, OR, and NOT. In a conventional electronic computer these are implemented by simple switching elements, or logic gates. For example, an AND gate has two input wires and one output wire (see figure 11). If an electrical pulse is received along both input wires, a pulse is sent out along the output wire. There is no output if only one or no input pulse is received. The computer consists of a very large network of such logic elements. Mathematics is performed by representing numbers in binary form, as strings of ones and zeros. Translated into physical form, a one is encoded as an electrical pulse, a zero as the absence of a pulse. There is, however, no need for these operations to be performed by electrical switching. Any device that executes the same logical operations will suffice. You could use mechanical gears (as in Charles Babbage's original Analytical Engine), laser beams, or dots on a computer screen.

Following much experimentation and thought, Conway was able to show that suitable logic circuits can indeed be built in the Life universe.

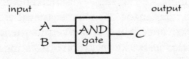

FIGURE 11. Symbolic representation of an AND gate used in computers. There are two input wires, A and B, and one output wire, C. If a signal is received along both A *and* B, then an output signal will be sent along C.

The essential idea is to use processions of gliders to code binary numbers. For example, the number 1011010010 can be represented by locating a glider in the procession in the position of each 1, while leaving gaps for the 0s. Logic gates can then be constructed by arranging glider streams to intersect at right angles in a controlled way. Thus an AND gate will emit a glider if and only if it simultaneously receives gliders from both input streams (thereby encoding the operation $1 + 1 \rightarrow 1$). To accomplish this, and to construct the necessary memory unit to store information, Conway needs just four Life species: gliders, glider guns, eaters, and blocks.

A lot of clever tricks are needed to position the elements correctly and orchestrate the dynamics. Nevertheless, the necessary logical circuitry can be organized, and the light shapes in the Life universe can function perfectly properly, if somewhat slowly, as a universal computer. It is a result with fascinating implications. There are two distinct levels of computation involved. First there is the underlying electronic computer used to produce the Life game on the screen; then the Life patterns themselves function as a computer on a higher level. In principle this hierarchy can continue indefinitely: the Life computer could be programmed to generate its own abstract Life universe, which in turn could be programmed to produce its Life universe. . . . Recently I attended a workshop on the study of complexity at which two MIT computer scientists, Tom Toffoli and Norman Margolus, demonstrated the operation of an AND gate on a computer monitor. Also watching the show was Charles Bennett of IBM, an expert on the mathematical foundations of computation and complexity. I remarked to Bennett that what we were watching was an electronic computer simulating a cellular automaton simulating a computer. Bennett replied that these successive embeddings of computational logic reminded him of Russian dolls.

The fact that Life can accommodate universal computers means that all the consequences of Turing's analysis can be imported into the Life universe. For example, the existence of uncomputable operations applies to the Life computer too. Remember that there is no systematic way to decide in advance whether a given mathematical problem is decidable or undecidable by the operation of a Turing machine: the fate of the machine cannot be known in advance. Therefore, the fate of the associated Life patterns cannot systematically be known in advance, even though all such patterns are strictly deterministic. I think this is a very profound conclusion, and one that has sweeping implications for the real world. It appears as if there is a kind of randomness or uncertainty (dare I call it "free will"?) built into the Life universe, as indeed there is in the real universe, *due to the restrictions of logic itself,* as soon as systems become complex enough to engage in self-reference.

Self-reference and self-reproduction are closely related, and once the existence of universal Life computers had been established, the way was open for Conway to prove the existence of universal constructors and hence genuine self-reproducing Life patterns. Again, no such pattern has actually been constructed, for it would be truly vast. But Conway reasons that, in an infinite Life universe randomly populated by dots, self-reproducing patterns would inevitably form *somewhere* just by chance. Although the odds against the spontaneous formation of such a complex and highly orchestrated pattern are astronomical, in a truly infinite universe anything that can happen will happen. One can even envisage Darwinian evolution leading to the appearance of ever more complex self-reproducing patterns.

Some Life enthusiasts assert that such self-reproducing Life patterns really would be alive, because they would possess all the attributes that define living organisms in our universe. If the essence of life is regarded simply as energy organized above a certain threshold of complexity, then they are right. In fact, there is now a distinct branch of science called "artificial life" which is devoted to studying self-organizing, adaptive, computer-generated patterns. The goal of the subject is to abstract the essence of what it means to be alive from the possibly irrelevant details of the actual materials of living organisms. At a recent workshop on artificial life, the computer scientist Chris Langton explained: "Our belief is that we can put sufficiently complex universes into computers so that they can support processes which, with respect to that universe, must be considered alive. But they wouldn't be made out of the same stuff. . . . It raises the awesome possibility that we are going to be creating the next living things in the universe."[6] Poundstone agrees: "If nontrivial self-reproduction is used as the criterion of life, then self-reproducing Life patterns would be alive. This is not to say that they would simulate life as any television image might, but that they would be literally alive by virtue of encoding and manipulating information about their own makeup. The simplest self-reproducing Life patterns would be alive in a sense that a virus is not."[7]

John Conway even goes so far as to suggest that advanced Life forms could be conscious: "It's probable, given a large enough Life space, initially in a random state, that after a long time, intelligent self-replicating animals will emerge and populate some parts of the space."[8] There is, however, a natural resistance to such ideas. The Life universe is, after all, only a simulated universe. It isn't real, is it? The shapes that move about on the screen only mimic real-life shapes. Their behavior isn't spontaneous, it is preprogrammed into the computer which plays the Life game. But, counters the Life enthusiast, the behavior of physical structures in our universe is also "programmed" by the laws of physics and the initial state. The random spread of dots from which a self-reproducing Life pattern might emerge is directly analogous to the random prebiotic soup from which the first living things are supposed to have emerged on Earth.

So how can we tell a real universe from a simulated one? That is the subject of the next chapter.

5

Real Worlds and Virtual Worlds

WE ARE ALL FASCINATED by dreams. Those people who, like myself, dream very vividly often have the experience of being "trapped" in a dream that we believe is real. The huge sense of relief that accompanies awakening is intensely genuine. Yet I have often wondered why, given that at the time the dream *is* the reality, we make such a sharp distinction between our experiences when awake and when asleep. Can we be absolutely sure that the "dream world" is illusory and the "awake world" real? Could it be the other way about, or that both are real, or neither? What criteria of reality can we employ to decide the matter?

The usual riposte is to claim that dreams are private experiences, whereas the world we perceive when awake is consistent with other people's experiences. But this is no help. I often encounter dream characters who assure me that they are real, and are sharing my own dream experiences. In waking life I have to take other people's word for it that they really do perceive a similar world to mine, because I cannot actually share their experiences. How am I able to distinguish a genuine claim from that made by an illusory character, or by a sufficiently complex, but unconscious, automaton? Nor is it any use pointing to the fact that dreams are often incoherent, fragmentary, and absurd. The so-called real world can often seem the same after a few glasses of wine, or when coming round from an anesthetic.

Simulating Reality

The above remarks about dreams are intended to soften the reader up for a discussion on computer simulations of reality. In the previous chapter I argued that a computer can simulate physical processes in the real world, in principle even those as complex as occur in biology. On the other hand, we saw that a computer is in essence simply a procedure for converting one set of symbols into another according to some rule. Usually we think of the symbols as numbers; more specifically as strings of ones and zeros, these being the most appropriate formulation for machines to use. Each one or zero represents a bit of information, so a computer is a device that takes an input-bit string and converts it into an output-bit string. How can this seemingly trivial set of abstract operations capture the essence of *physical* reality?

Compare the activity of the computer with a natural physical system—for example, a planet going round the sun. The state of the system at any instant can be specified by giving the position and velocity of the planet. These are the input data. The relevant numbers can be given in binary arithmetic, as a bit string of ones and zeros. At some later time the planet will have a new position and velocity, which can be described by another bit string: these are the output data. The planet has succeeded in converting one bit string into another, and is therefore in a sense a computer. The "program" it has used in this conversion is a set of physical laws (Newton's laws of motion and gravitation).

Scientists are becoming increasingly aware of the link between physical processes and computation, and are finding it profitable to think of the world in computational terms. "Scientific laws are now being viewed as algorithms," according to Stephen Wolfram of the Institute for Advanced Study in Princeton. "Physical systems are viewed as computational systems, processing information much the way computers do."[1] Consider a gas, for example. The state of the gas is specified by giving the positions and velocities of all the molecules at an instant (to some accuracy). This would be an enormously long bit string. At a later instant the state of the gas would define another enormously long bit string. The effect of the dynamical evolution of the gas has therefore been to convert input data into output data.

The connection between natural processes and computational operations is further strengthened by the quantum theory, which reveals that many physical quantities normally regarded as continuous are in fact discrete. Thus atoms possess distinct energy levels. When an atom changes its energy, it makes a jump between levels. If each level is assigned a number, then such a jump can be considered as a transition from one number to another.

We have here arrived at the essence of the efficacy of the computer in modern science. Because of the ability for computers to simulate one another, an electronic computer is capable of simulating any system that itself acts like a computer. This is the basis for computer modeling of the real world: planets and boxes of gas and much else do act like computers and so can be modeled. But can every physical system be simulated this way? Wolfram thinks so: "One expects the fact that computers are as powerful in their computational capabilities as any physically realizable system can be, so that they can simulate any physical system."[2] If true, this implies that any system complex enough to compute can in principle simulate *the entire physical universe*.

In the previous chapter I explained how cellular automata such as Life generate toy universes in which computation is possible. We seem to have arrived at the conclusion that the Life universe is capable of faithfully mimicking the real universe. "Cellular automata that are capable of universal computation can mimic the behavior of any possible computer," explains Wolfram. So that, "since any physical process can be represented as a computational process, they can mimic the action of any possible physical system as well."[3] So, could a cellular-automaton toy universe, like the Life universe, in principle be made so "lifelike" that it could serve as a perfect replica of the real universe? Apparently so. But this raises a further perplexing question. If all physical systems are computers, and if computers can perfectly mimic all physical systems, then what distinguishes the real world from a simulation?

One is tempted to answer that simulations are only imperfect approximations to reality. When the motion of a planet is calculated, for example, the accuracy of the input data is limited by observation. Furthermore, realistic computer programs greatly simplify the physical situation, by neglecting the disturbing effect of minor bodies and so on. But one can certainly *imagine* more and more refined programs, and more and more elaborate data-gathering, until the simulation is, for all practical purposes, indistinguishable from the reality.

But wouldn't the simulation necessarily fail at some level of detail? For a long time it was believed that the answer must be yes, on account of what was assumed to be a fundamental difference between real physics and any digital simulation. This difference has to do with the question of time-reversibility. As explained in chapter 1, the laws of physics are reversible in the sense that they remain unchanged if past and future are inverted—i.e., they have no inbuilt preferred time direction. Now, all existing digital computers expend energy to operate. This wasted energy appears as heat inside the machine and has to be got rid of. The accumulation of heat imposes very practical limitations on the performance of computing machines, and a lot of research goes into minimizing it. The difficulty can be traced to the essential logical elements in the computer. Whenever switching occurs, heat is produced. This is familiar from daily life. The clunk you hear when you throw a light switch is some of the energy you expended in throwing the switch being dissipated as sound; the rest appears as heat inside the switch. This energy cost is deliberately incorporated in the design of the switch to ensure it remains in one or the other of its two stable states—on or off. If there were no energy cost involved in switching, there would be a danger that the switch might flip spontaneously.

The energy dissipation in switching is irreversible. The heat flows away into the environment and is lost. You can't scoop up wasted heat energy and channel it back into something useful without incurring a further heat loss at least as great in the scooping-up process. This is an example of the second law of thermodynamics, which forbids any "free-lunch" recycling of heat energy for useful work. It was realized by some computer scientists, however, that the second law of thermodynamics is a statistical law which applies to systems with many degrees of freedom. Indeed, the very notions of heat and entropy involve the chaotic agitation of molecules, and are meaningful for large collections of molecules only. If computers could be miniaturized so much that the basic switching were conducted at the molecular level, could the heat generation perhaps be avoided altogether?

However, there seemed to be a basic principle that contradicted the existence of this idealization. Consider the AND gate, for example, described in the previous chapter. The input has two channels (wires), the output only one. The whole purpose of the AND operation is to merge two incoming signals into a single outgoing signal. Obviously this is not reversible. You can't tell whether the absence of a pulse in the out wire is due to there having been a pulse in only one input wire, or the other, or no pulse in either. This elementary limitation reflects the obvious fact that in ordinary arithmetic you can deduce answers from questions, but not the other way about: you generally can't deduce the question from the answer. If you are told the answer to a sum is 4, the sum might have been 2 + 2 or 3 + 1 or 4 + 0. It might appear from this that no computer could be run backward for reasons of basic logic.

In fact, there is a flaw in this argument, which was recently uncovered by Rolf Landauer and Charles Bennett of IBM. They traced back the irreversibility that is seemingly inherent in computing and showed that it stems from throwing information away. Thus, performing the sum 1 + 2 + 2, one might first add 2 and 2 to get 4, then add 4 to 1 to obtain the answer 5. In this sequence of operations there is an intermediate step at which only the number 4 is retained: the original 2 + 2 is discarded as no longer relevant to the remaining part of the calculation. But we do not have to throw the information away. We could choose to keep track of it throughout. Of course, this would mean enlarging the memory space to accommodate the extra information, but it would enable us to "undo" any calculation at any stage by working backward from the answer to the question.

But can suitable switching gates be designed to implement this reversible logic? Indeed they can, as discovered by Ed Fredkin of MIT. The Fredkin gate has two input and two output channels plus a third "control channel." Switching is achieved as usual, but in a way that retains the input information in the output channels. A computation can be conducted reversibly even on a dissipative machine—i.e., on one that inevitably dissipates energy irreversibly. (Any practical reversible computation could not avoid the irreversible dissipation of heat.) But at the theoretical level one can imagine an idealized system in which both the computation and the physics would be reversible. Fredkin has devised an imaginary arrangement of rigid balls bouncing in a carefully supervised way from immovable baffles. This setup can actually implement logical operations reversibly. Other imaginary reversible computers have also been concocted.

An interesting question concerns the status of cellular automata as computers. Life-game computers are not reversible, because the underlying rules for the game are not reversible (the pattern sequences can't be run backward). However, a different type of cellular automaton which can model the reversible Fredkin ball-and-baffle system has been constructed by Norman Margolus. Viewed at the level of the automaton universe, this is a genuinely reversible computer, both computationally and "physically" (although there is still irreversible dissipation at the level of the electronic computer that implements this cellular automaton).

The fact that computation can be carried out reversibly removes a crucial distinction between a computer simulation and the real-world physics it is simulating. Indeed, one can invert the reasoning and ask to what extent real-world physical processes *are* computational processes. If irreversible switches are unnecessary, can the movement of ordinary bodies be regarded as part of a digital computation? Some years ago it was proved that certain irreversible systems, like Turing machines and cellular automata with noninvertible rules such as Life, can be programmed to perform any digital computation whatever by appropriately choosing their initial state. This property is called "computational universality." In the case of Life it implies that an initial pattern could be chosen which would place a dot in a given location if, say, a certain number is prime. Another pattern would do so if a certain equation has a solution, and so on. In this way, the Life game could be used to investigate unsolved mathematical problems like Fermat's last theorem.

More recently it has been shown that certain reversible deterministic systems such as Fredkin's ball-and-baffle computer are also computationally universal, and that even some nondeterministic systems share this property. It therefore seems as if computational universality is a fairly common property of physical systems. If a system does have this property, it is by definition capable of behavior as complex as any that can be digitally simulated. There is evidence that even a system as simple as three bodies moving under mutual attraction (e.g., two planets orbiting a star) possesses the property of computational universality. If so, then by appropriately choosing the positions and velocities of the planets at one instant, the system could be made to compute, say, the digits of pi, or the trillionth prime, or the outcome of a billion-glider collision in the Life universe. Indeed, this seemingly trivial trinity could even be used to simulate the whole universe if, as some enthusiasts claim, the universe is digitally simulatable.

We are used to thinking of computers as very special systems that require ingenious design. Certainly electronic computers are complicated, but this is because they are very versatile. A lot of the programming work is already taken care of in the design of the machine: we do not have to redo it every time through the initial conditions. But the ability to compute is something that many physical systems, including some very simple ones, seem to possess. This raises the question of whether the activities of atoms or even subatomic particles can compute. A study of this point was made by the physicist Richard Feynman, who showed that a reversible computer operating at the subatomic level in accordance with the laws of quantum mechanics is indeed a possibility. So can we regard the countless atomic processes going on quite naturally all the time—processes inside you and me, inside the stars, the interstellar gas, distant galaxies—as part of some gigantic cosmic computation? If so, then physics and computation become identical and we would arrive at an astonishing conclusion: the universe would be its own simulation.

Is the Universe a Computer?

One person who answers emphatically yes to this question is Ed Fredkin. He believes that the physical world is a gigantic cellular automaton, and claims that the study of cellular automata is revealing that realistic physical behavior, including refinements such as relativity, can be simulated. Fredkin's colleague Tom Toffoli shares this belief. He once quipped that of course the universe is a computer, the only trouble is that somebody else is using it. And as for us, well, we are just bugs in the great cosmic machine! "All we have to do," he claims, "is 'hitch a ride' on this huge ongoing computation, and try to discover which parts of it happen to go near where we want."[4]

Fredkin and Toffoli are not without supporters for this startling—one might even say bizarre—viewpoint. The physicist Frank Tipler has also argued strongly for the idea of equating the universe with its own simulation. Furthermore, the simulation need not be run on an actual computer, Tipler maintains. A computer program is, after all, just a conversion (or mapping) of one set of abstract symbols into another according to some rule: input → output. A physical computer provides a concrete representation of such a mapping, just as the Roman numeral III is a representation of the abstract number three. The mere existence of such a mapping—even abstractly, in the realm of mathematical rules—is enough for Tipler.

123

It has to be pointed out that our present theories of physics are not generally formulated in quite the same way as computer algorithms, because they make use of quantities that vary continuously. In particular, space and time are considered to be continuous. "The possibility that there is to be an exact simulation, that the computer will do exactly the same as nature," explains Richard Feynman, demands "that everything that happens in a finite volume of space and time would have to be exactly analyzable with a finite number of logical operations. The present theory of physics is not that way, apparently. It allows space to go down to infinitesimal distances."[5] On the other hand, the continuity of space and time are only assumptions about the world. They cannot be proved, because we can never be sure that at some small scale of size, well below what can be observed, space and time might not be discrete. What would this mean? For one thing it would mean that time advanced in little hops, as in a cellular automaton, rather than smoothly. The situation would resemble a movie film which advances one frame at a time. The film appears to us as continuous, because we cannot resolve the short time intervals between frames. Similarly, in physics, our current experiments can measure intervals of time as short as 10^{-26} seconds; there are no signs of any jumps at that level. But, however fine our resolution becomes, there is still the possibility that the little hops are yet smaller. Similar remarks apply to the assumed continuity of space. So this objection to an exact simulation of reality is perhaps not fatal.

One is nevertheless still tempted to object that the map is distinct from the territory. Even if there could exist a cosmic computer so unthinkably powerful that it would be capable of simulating exactly the activity of every atom in the universe, surely the computer doesn't actually contain a planet Earth moving in space, any more than the bible contains Adam and Eve? A computer simulation is usually regarded as just a representation, or an image, of reality. How could anybody claim that the activity going on inside an electronic computer could ever create a real world?

Tipler counters that this objection is valid only from a perspective outside the computer. If the computer were powerful enough to simulate consciousness—and, by extension, a whole community of conscious beings—from the viewpoint of the beings *within* the computer the simulated world would be *real*:

The key question is this: do the simulated people exist? As far as the simulated people can tell, they do. By assumption, any action which real people can and do carry out to determine if they exist—reflecting on the fact that they think, interacting with the environment—the simulated people also can do, and in fact do do. There is simply no way for the simulated people to tell that they are "really" inside the computer, that they are merely simulated, and not real. They can't get at the real substance, the physical computer, from where they are, inside the program. . . . There is no way for the people inside this simulated universe to tell that they are merely simulated, that they are only a sequence of numbers being tossed around inside a computer, and are in fact not real.[6]

Of course, Tipler's entire discussion hinges on the possibility that a computer can simulate consciousness. Is this reasonable? Imagine the computer simulating a human being. If the simulation were truly accurate, an external human observer who did not know the circumstances would be unable to tell by conversing with the simulation whether it was located in the computer or was a human being in our world. The observer could interrogate the simulation and obtain perfectly sensible, humanlike answers. As a result the observer would be tempted to conclude that the simulation was genuinely conscious. In fact, Alan Turing himself addressed this issue in a famous paper entitled "Can Machines Think?" in which he devised just such an interrogation test. Even though most people regard the idea of machines possessing consciousness as outlandish or even absurd, many distinguished scientists and philosophers of the so-called strong-AI school have argued on this basis that a simulated mind would be conscious.

For those prepared to go along with the idea that a sufficiently powerful computer could be conscious, it is but a small step to accept that a computer could, in principle, generate an entire society of conscious beings. These individuals would presumably think, feel, live, and die in their simulated world entirely oblivious of the fact that they exist by courtesy of some computer operator who could presumably pull the plug at any moment! This would be precisely the position of Conway's intelligent animals in the Life universe.

But this whole discussion begs the obvious question: how do we know that we ourselves are "real," and not merely a simulation inside a gigantic computer? "Obviously, we can't know," says Tipler. But does it matter? Tipler argues that the actual existence of the computer, being unverifiable to the conscious beings within it, is irrelevant. All that matters is the existence of a suitable abstract program (even an abstract look-up table would do) capable of simulating a universe. By the same token, the actual existence of a physical universe is irrelevant: "Such a physically real universe would be equivalent to a Kantian thing-in-itself. As empiricists, we are forced to dispense with such an inherently unknowable object: the universe must be an abstract program."[7]

The drawback with this position (quite apart from its air of *reductio ad absurdum*) is that the number of possible abstract programs is infinite. Why do we experience this particular universe? Tipler believes that all possible universes that can support consciousness *are* actually experienced. Ours is not the only one. Obviously we see this one, by definition. But other universes exist, many of them similar to ours, with their own inhabitants, for whom their universe is to them every bit as real as ours is to us. (This is a variant of the "many-universes" interpretation of quantum mechanics, popular with a lot of distinguished physicists, described in detail in my book *Other Worlds*. I shall return to it in chapter 8.) Those programs that encode universes incapable of supporting conscious beings go unobserved, and perhaps can therefore be regarded as in some sense less than real. The set of programs capable of generating cognizable universes will be a small subset of the set of all possible programs. Ours can be regarded as typical.

The Unattainable

If the universe is the "output" of some computational process, then it must be, by definition, computable. More precisely, there must exist a program or an algorithm from which a correct description of the world may be obtained in a finite number of steps. If we knew that algorithm, we would have a complete theory of the universe, including the numerical values for all measurable physical quantities. What can one say about these numbers? If they are to emerge from a computation, they will have to be *computable* numbers. It has generally been assumed that the values of all measurable quantities predicted by physical theory would be computable numbers. But recently this assumption has been challenged by physicists Robert Geroch and James Hartle. They point out that existing theories of physics might yield predictions for measurable quantities that are uncomputable numbers. Though these theories have to do with the rather technical subject of the quantum properties of space-time, they raise an important point of principle.

Suppose a cherished theory predicts an uncomputable number x for some quantity—for example, the ratio of two subatomic-particle masses. Can the theory be tested? Testing any prediction involves comparing the theoretical value with the experimental value. Obviously this can be done only to within some level of accuracy. Suppose the experimental value is determined to within an expected error of 10 percent. It would then be necessary to know x to within 10 percent. Now, although x may exist, no finite algorithm, no systematic procedure, can find it; that is what is meant by its being uncomputable. On the other hand, we need to know x only to within 10 percent. It is certainly possible to find an algorithm to produce a sequence of successively better approximations to x, eventually to within 10 percent. The trouble is, as we don't know x, we can't know when we've got to the 10-percent level.

In spite of these difficulties, it might be possible to find a 10-percent approximation by nonalgorithmic means. The point about an algorithmic construction is that one can lay down a finite set of standard instructions at the outset; it is then purely a mechanical matter to work through those instructions to obtain the desired result. In the case of a computable number such as pi, one can imagine a computer churning away, generating a sequence of ever better approximations and including in the output at each step precisely how good that particular approximation was. But as we have seen, this strategy won't work for an uncomputable number. Instead the theorist would have to approach each level of accuracy as a new problem, to be tackled a different way. By some clever trick it might be possible to find a 10-percent approximation to x. But the same trick wouldn't necessarily work to reach the 1-percent level. The theorist would be obliged to try some totally different strategy. With each improvement in experimental accuracy the poor theorist would need to work harder and harder to find a matching approximation for the predicted value.

As Geroch and Hartle point out, finding a theory is normally the hard part; implementing it is usually a purely mechanical procedure. It took Newton's genius to come up with the laws of motion and gravitation, but a computer can be programmed to implement the theory "blindly" and predict the date for the next solar eclipse. In the case of a theory which predicts uncomputable numbers, implementing the theory may be just as hard as finding it in the first place. Indeed, no clear distinction can be drawn between these two activities.

It would obviously be better for the theorist if our physical theories were never like this. However, we cannot be sure that they will always be so. There may be compelling reasons for a particular theory, which then turns out to yield uncomputable predictions, as Geroch and Hartle suggest may be the case for the quantum description of space-time. Should the theory be discarded solely on that account? Is there any reason why the universe has to be "algorithmically implementable"? We just don't know, but one thing is sure. If it is not, the otherwise very close analogy between nature and the computer breaks down.

Following Einstein's dictum that God is subtle but not malicious, let us assume that we do indeed live in a "computable" universe. What, then, can we learn about the nature of the program that the likes of Fredkin and Tipler would have us believe is the source of our reality?

The Unknowable

Consider for a moment the case of a program used in an electronic computer—for example, to multiply a string of numbers. The essence of the concept is that the program should be in some sense easier to construct than the operations it is intended for. If this were not so, one wouldn't bother with the computer, but would simply carry out the arithmetic operations directly. One way of putting this is to say that a useful computer program can generate more information (in the example, the results of many multiplications) than it contains itself. This is no more than a fancy way of saying that in mathematics we look for simple rules that can be used again and again, even in very complicated calculations. However, not all mathematical operations can be carried out by a program significantly less complicated than the operation itself. Indeed, the existence of uncomputable numbers implies that for some operations there exists *no* program. Thus some mathematical processes are intrinsically so complex that they cannot be encapsulated in a compact program at all.

In the natural world we are also faced by enormous complexity, and the question arises as to whether a description of this complexity can be captured in a compact description. Put differently, is the "program for the universe" significantly simpler than the universe itself? This is a very deep question about the nature of the physical world. If a computer program or an algorithm is simpler than the system it describes, the system is said to be "algorithmically compressible." So we are faced by the question of whether the universe is algorithmically compressible.

Before we turn to this question, it will be helpful to consider the idea of algorithmic compression in a little more detail. The subject of algorithmic information theory was created in the 1960s in the Soviet Union by Andrei Kolmogorov and in the United States by Gregory Chaitin of IBM. The essence of the idea hinges on a very simple question: what is the shortest message that can describe a system to a certain level of detail? Obviously a simple system can be described easily, but a complex system can't. (Try describing the structure of a coral reef in the same number of words needed to describe an ice cube.) Chaitin and Kolmogorov suggest that the complexity of something is defined as the length of the shortest possible description of that thing.

Let's see how this works for numbers. There are simple numbers, such as 2 or pi, and complicated ones, such as a string of 1s and 0s generated by coin tossing (heads = 0, tails = 1). What sort of descriptions can we give that will define such numbers uniquely? One strategy is simply to write them out in decimal or binary form (pi can only be given to a particular approximation, because it has an infinite decimal expansion). But this is not, obviously, the most economical description. The number pi, for example, would be better described by giving a formula that can be used to calculate it to any desired degree of approximation. If the numbers concerned are regarded as the output of a computer, then the shortest description of a number will be the shortest program that will enable the computer to output that number. Simple numbers will be generated by short programs, complex numbers by long programs.

The next step is to compare the length of the number with the length of the program that generates it. Is it shorter? Has a compression been achieved? To make this more precise, suppose that the output of the computer is expressed as a string of ones and zeros, such as

$$101101011100010100110101001 \ldots$$

(where ". . ." denotes "and so on, perhaps forever"). This string will have a certain information content, which is measured in "bits." We want to compare the amount of information in the output with the information content of the program itself. To take a simple example, suppose the output is

101010101010101010101010101010

This could be generated by the simple algorithm "Print 01 fifteen times." A very much longer output string could be generated by the program "Print 01 one million times." The second program is scarcely more complicated than the first, yet it produces vastly greater output information. The lesson is that, if the output contains any patterns, then these may be compactly encoded in a simple algorithm that can be very much shorter (in terms of bits of information) than the output itself. In this case the string is said to be algorithmically compressible. If, conversely, a string cannot be generated by an algorithm significantly shorter than itself, it is algorithmically incompressible. In this case the string will possess no regularities or patterns whatsoever. It will just be a haphazard collection of 1s and 0s. In this way the amount of algorithmic compression attainable can be seen to be a useful measure of the simplicity or structure present in the output, with low compressibility being a measure of complexity. Simple, regular strings are highly compressible, whereas complex, patternless ones are less so.

Algorithmic compression provides a rigorous definition of randomness: a random sequence is one that cannot be algorithmically compressed. It may not be easy to tell merely by looking whether a given string is compressible. It might possess patterns of great subtlety built into it in a cryptic way. Every code-breaker knows that what looks at a glance like a random jumble of letters might in fact be a structured message; to tell, all you need is the code. The infinite decimal expansion (and its binary counterpart) of the number pi shows no obvious patterns at all over thousands of digits. The distribution of digits passes all the standard statistical tests for randomness. From a knowledge of the first thousand digits alone there is no way of predicting what the thousand-and-first will be. Yet in spite of this pi is *not* algorithmically random, because a very compact algorithm can be written to generate its expansion.

Chaitin points out that these ideas of mathematical complexity can be convincingly extended to physical systems: the complexity of a physical system is the length of the minimal algorithm that can simulate or describe it. At first sight this approach seems to be rather arbitrary, because we have not yet specified which make of computer is to be used. It turns out, however, that it doesn't really matter, because all universal computers can simulate one another. Similarly, the computer language we choose to work in—LISP, BASIC, FORTRAN—is irrelevant. It is a straightforward matter to write instructions to translate one computer language into another. It turns out that the extra program-length needed to convert the language and to run the program on another machine is typically a very small correction to the total program-length. So you don't have to worry about how the computer you use is actually made. This is an important point. The fact that the definition of complexity is machine-independent suggests that it captures a really existing quality of the system, and is not merely a function of the way we choose to describe it.

A more legitimate concern is how one can know whether any particular algorithm is actually the shortest possible. If a shorter one is found, then the answer is clearly no. But it turns out to be impossible in general to be sure that the answer is yes. The reason can be traced back to Gödel's theorem on undecidability. Remember that this theorem was based on a mathematical version of the "liar" self-reference paradox ("This statement is false"). Chaitin adapted the idea to statements about computer programs. Consider the case that a computer is given the following command: "Search for a string of digits that can only be generated by a program longer than this one." If the search succeeds, the search program itself will have generated the digit string. But then the digit string cannot be "one that can only be generated by a program longer than this." The conclusion must be that the search will fail, even if it continues forever. So what does that tell us? The search was intended to find a digit string that needed a generating program at least as big as the search program, which is to say that any shorter program is ruled out. But as the search fails, we cannot rule out a shorter program. We simply don't know in general whether a given digit string can be encoded in a program shorter than the one we happen to have discovered.

Chaitin's theorem has an interesting implication for random number sequences—i.e., random digit strings. As explained, a random sequence is one that can't be algorithmically compressed. But as we have just seen, you cannot know whether or not a shorter program exists for generating that sequence. You can never tell whether you have discovered all the tricks to shorten the description. So you can't in general prove that a sequence is random, although you could disprove it by actually finding a compression. This result is all the more curious since it can be proved that almost all digit strings are random. It is just that you cannot know precisely which!

It is fascinating to speculate that apparently random events in nature may not be random at all according to this definition. We cannot be sure, for example, that the indeterminism of quantum mechanics may not be like this. After all, Chaitin's theorem ensures we can never prove that the outcome of a sequence of quantum-mechanical measurements is actually random. It certainly *appears* random, but so do the digits of pi. Unless you have the "code" or algorithm that reveals the underlying order, you might as well be dealing with something that is truly random. Could there exist a more elaborate sort of "cosmic code," an algorithm that would generate the results of quantum events in the physical world and hence expose quantum indeterminism as an illusion? Might there be a "message" in this code that contains some profound secrets of the universe? This idea has already been seized on by some theologians, who have noticed that quantum indeterminism offers a window for God to act in the universe, to manipulate at the atomic level by "loading the quantum dice," without violating the laws of classical (i.e., nonquantum) physics. In this way God's purposes could be imprinted on a malleable cosmos without upsetting the physicists too much. In chapter 9 I shall describe a specific proposal of this sort.

Armed with his algorithmic definition, Chaitin has been able to demonstrate that randomness pervades all mathematics, including arithmetic. To do this he has discovered a monstrous equation containing seventeen thousand variables (technically known as a Diophantine equation). The equation contains a parameter K which can take integer values 1,2,3, and so on. Chaitin now asks whether, for a given value of K, his monster equation has a finite or an infinite number of solutions. One can imagine plodding in turn through every value of K, recording the answers: "finite," "finite," "infinite," "finite," "infinite," "infinite." . . . Will there be any pattern to this sequence of answers? Chaitin has proved that there will be no pattern. If we represent "finite" by 0 and "infinite" by 1, then the resulting digit string 001011 . . . cannot be algorithmically compressed. It will be random.

The implications of this result are startling. It means that, in general, if you pick a value of K, there is no way of knowing without explicitly checking whether or not that particular Diophantine equation possesses a finite or an infinite number of solutions. In other words, there is no systematic procedure for deciding in advance the answers to perfectly well-defined mathematical questions: the answers are random. Nor can solace be sought in the fact that a seventeen-thousand-variable Diophantine equation is a rather special mathematical oddment. Once randomness has entered mathematics it infests it through and through. The popular image of mathematics as a collection of precise facts, linked together by well-defined logical paths, is revealed to be false. There is randomness and hence uncertainty in mathematics, just as there is in physics. According to Chaitin, God not only plays dice in quantum mechanics, but even with the whole numbers. Chaitin believes that mathematics will have to be treated more like the natural sciences, in which results depend on a mixture of logic and empirical discovery. One might even foresee universities with departments of experimental mathematics.

An amusing application of algorithmic information theory concerns an uncomputable number known as omega, which Chaitin defines to be the probability that a computer program will halt if its input consists merely of a random string of binary numbers. The probability of something is a number between 0 and 1: the value 0 corresponds to the thing being impossible, the value 1 to its being inevitable. Obviously omega will be close to 1, because most random inputs will appear as garbage to the computer, which will then rapidly halt, displaying an error message. However, it can be shown that omega is algorithmically incompressible, and its binary or decimal expansion is completely random after the first few digits. Because omega is defined by reference to the halting problem, it encodes a solution to the halting problem in the sequence of its digits. Thus the first n digits in the binary expansion of omega will contain the answer to the problem of which n-digit programs will halt and which will run forever.

Charles Bennett has pointed out that many of the outstanding unsolved problems of mathematics, such as Fermat's last theorem, can be formulated as a halting problem, because they consist of conjectures that something does not exist (in this case a set of numbers that satisfies Fermat's theorem). The computer merely needs to search for a counterexample. If it finds one, it will halt; if not, it will chug on forever. Moreover, most interesting problems could be encoded in programs of only a few thousand digits' length. So knowing merely the first few thousand digits of omega would give us access to a solution of all outstanding mathematical problems of this type, as well as all other problems of similar complexity that might be formulated in the future! "It embodies an enormous amount of wisdom in a very small space," writes Bennett, "inasmuch as its first few thousand digits, which could be written on a small piece of paper, contain the answers to more mathematical questions than could be written down in the entire universe."[8]

Unfortunately, being an uncomputable number, omega can never be revealed by constructive means, however long we work at it. Thus, short of a mystical revelation, omega can never be known to us. And even if we were to be given omega by divine transmission, we would not recognize it for what it was, because, being a random number, it would not commend itself to us as special in any respect. It would be just a jumble of patternless digits. For all we know, a significant piece of omega could be written down in a textbook somewhere.

The wisdom contained in omega is real, but forever hidden from us by the strictures of logic and the paradoxes of self-reference. Omega the Unknowable is perhaps the latter-day counterpart of the "magic numbers" of the ancient Greeks. Bennett is positively poetic about its mystical significance:

Throughout history philosophers and mystics have sought a compact key to universal wisdom, a finite formula or text which, when known and understood, would provide the answer to every question. The Bible, the Koran, the mythical secret books of Hermes Trismegistus, and the medieval Jewish Cabala have been so regarded. Sources of universal wisdom are traditionally protected from casual use by being hard to find, hard to understand when found, and dangerous to use, tending to answer more and deeper questions than the user wishes to ask. Like God the esoteric book is simple yet undescribable, omniscient, and transforms all who know it. . . . Omega is in many senses a cabalistic number. It can be known of, but not known, through human reason. To know it in detail, one would have to accept its uncomputable digit sequence on faith, like words of a sacred text.[9]

The Cosmic Program

Algorithmic information theory provides a rigorous definition of complexity based on the ideas of computation. Pursuing our theme of the universe as a computer—or, more accurately, a computation—the question arises as to whether the immense complexity of the universe is algorithmically compressible. Is there a compact program that can "generate" the universe in all its intricate detail?

Although the universe is complex, it is clearly not random. We observe regularities. The sun rises each day on schedule, light always travels at the same speed, a collection of muons always decay with a half-life of two-millionths of a second, and so on. These regularities are systematized into what we call laws. As I have already stressed, the laws of physics are analogous to computer programs. Given the initial state of a system (input), we can use the laws to compute a later state (output).

The information content of the laws plus initial conditions is generally far less than that of the potential output. Of course, a law of physics may look simple when written down on paper, but it is usually formulated in terms of abstract mathematics, which itself needs a bit of decoding. Still, the information required to understand the mathematical symbols is limited to a few textbooks, whereas the number of facts described by these theories is unlimited. A classic example is provided by the prediction of eclipses. Knowing the position and motion of the Earth, sun, and moon at one time enables us to predict the dates of future (and past) eclipses. So one set of input data yields many output sets. In computer jargon, we may say that the data set of eclipses has been algorithmically compressed into the laws plus initial conditions. Thus the observed regularities of the universe are an example of its algorithmic compressibility. Underlying the complexity of nature is the simplicity of physics.

Interestingly, one of the founders of algorithmic information theory, Ray Solomonoff, concerned himself with precisely these sorts of questions. Solomonoff wanted to find a way of measuring the relative plausibility of competing scientific hypotheses. If a given set of facts about the world can be explained by more than one theory, how do we choose between them? Can we assign some sort of quantitative "value" to the contenders? The short answer is to use Occam's razor: you pick the theory with the least number of independent assumptions. Now, if one thinks of a theory as a computer program, and the facts of nature as the output of that program, then Occam's razor obliges us to pick the shortest program that can generate that particular output. That is to say, we should prefer the theory, or program, which offers the greatest algorithmic compression of the facts.

Viewed this way, the entire scientific enterprise can be seen as the search for algorithmic compressions of observational data. The goal of science is, after all, the production of an abbreviated description of the world based on certain unifying principles we call laws. "Without the development of algorithmic compressions of data," writes Barrow, "all science would be replaced by mindless stamp collecting—the indiscriminate accumulation of every available fact. Science is predicated upon the belief that the Universe is algorithmically compressible and the modern search for a Theory of Everything is the ultimate expression of that belief, a belief that there is an abbreviated representation of the logic behind the Universe's properties that can be written down in finite form by human beings."[10]

So can we conclude that cosmic complexity can all be compressed into a very short "cosmic program," much the way complexity in the Life universe boils down to a simple set of rules repeatedly applied? Although there are many conspicuous examples of algorithmic compression in nature, not every system can be thus compressed. There is a class of processes, the importance of which has only recently been recognized, known as "chaotic." These are processes that exhibit no regularities. Their behavior appears to be entirely random. Consequently they are not algorithmically compressible. It used to be thought that chaos was rather exceptional, but scientists are coming to accept that very many natural systems are chaotic, or can become so under certain circumstances. Some familiar examples include turbulent fluids, dripping taps, fibrillating hearts, and driven pendula.

Even though chaos is rather common, it is clear that on the whole the universe is far from being random. We recognize patterns everywhere and codify them into laws which have real predictive power. But the universe is also far from being simple. It possesses a subtle kind of complexity that places it partway between simplicity on the one hand and randomness on the other. One way of expressing this quality is to say that the universe has "organized complexity," a topic that I have discussed at length in my book *The Cosmic Blueprint*. There have been many attempts to capture mathematically this elusive element called organization. One is due to Charles Bennett, and involves something he calls "logical depth." This focuses less on the quantity of complexity or the amount of information needed to specify a system, and more on its quality, or "value." Bennett explains:

> A typical sequence of coin tosses has high information content, but little message value; an ephemeris, giving the positions of the moon and planets every day for a hundred years, has no more information than the equations of motion and the initial conditions from which it was calculated, but saves its owner the effort of recalculating these positions. The value of a message thus appears to reside . . . in what might be called its buried redundance—parts predictable only with difficulty, things the receiver could in principle have figured out without being told, but only at considerable cost in time, money and computation. In other words, the value of a message is the amount of mathematical or other work plausibly done by its originator, which its receiver is saved from having to repeat.[11]

Bennett invites us to think about the state of the world as having coded information folded up in it, information about the way the state was achieved in the first place. The issue is then how much "work" the system had to do—i.e., how much information processing went on—to reach that state. This is what he refers to as logical depth. The amount of work is made precise by defining it in terms of the time taken to compute the message from the shortest program that will generate it. Whereas algorithmic complexity focuses on the length of the minimal program to yield a given output, logical depth is concerned with the running time for the minimal program to generate that output.

Of course, you can't tell just by looking at some computer output precisely how it was produced. Even a quite detailed and meaningful message might have been produced by random processes. In that well-worn example, given time a monkey will type the works of Shakespeare. But according to the ideas of algorithmic information theory (and Occam's razor), the most plausible explanation for the output is to identify its cause with the minimal program, because that involves the least number of *ad hoc* assumptions.

Put yourself in the position of a radio astronomer who picks up a mysterious signal. The pulses, when arranged in sequence, are the first million digits of pi. What are you to conclude? Belief that the signal is random involves a million bits' worth of *ad hoc* assumptions, whereas the alternative explanation—that the message originated with some mechanism programmed to compute pi—would be much more plausible. In fact, a real episode of this sort occurred in the 1960s, when Jocelyn Bell, a Cambridge Ph.D. student working with Anthony Hewish on radio astronomy, picked up regular pulses from an unknown source. However, Bell and Hewish soon rejected the hypothesis that the pulses were artificial. Unlike the digits of pi, a series of precisely spaced pulses has little logical depth—it is logically shallow. There are many plausible explanations with few *ad hoc* assumptions for such a regular pattern, because many natural phenomena are periodic. In this case the source was soon identified with a rotating neutron star, or pulsar.

Simple patterns are logically shallow, because they may be generated rapidly by short and simple programs. Random patterns are also shallow, because their minimal program is, by definition, not much shorter than the pattern itself, so again the program is very short and simple: it need only say something like "Print pattern." But highly organized patterns are logically deep, because they require that many complicated steps be performed in generating them.

One obvious application of logical depth is to biological systems, which provide the most conspicuous examples of organized complexity. A living organism has great logical depth, because it could not plausibly have originated except through a very long and complicated chain of evolutionary processes. Another example of a deep system can be found in the complex patterns generated by cellular automata such as Life. In all cases the rule used is very simple, so from the algorithmic point of view these patterns actually have low complexity. The essence of Life's complexity lies not with the rules, but with their repeated use. The computer has to work very hard applying the rule again and again before it can generate deeply complex patterns from simple initial states.

The world abounds in deep systems, which show evidence of enormous "work" in fashioning them. Murray Gell-Mann once remarked to me that deep systems can be recognized because they are the ones we want to preserve. Shallow things can easily be reconstructed. We value paintings, scientific theories, works of music and literature, rare birds, and diamonds because they are all hard to manufacture. Motorcars, salt crystals, and tin cans we value less; they are relatively shallow.

So what can we conclude about the cosmic program? For centuries scientists have loosely talked about the universe being "ordered," without having a clear distinction between the various types of order: simple and complex. The study of computation has enabled us to recognize that the world is ordered both in the sense of being algorithmically compressible, and in the sense of having depth. The order of the cosmos is more than mere regimented regularity, it is also organized complexity, and it is from the latter that the universe derives its openness and permits the existence of human beings with free will. For three hundred years science has been dominated by the former: the search for simple patterns in nature. In recent years, with the advent of fast electronic computers, the truly fundamental nature of complexity has been appreciated. So we see that the laws of physics have a twofold job. They must provide the simple patterns that underlie all physical phenomena, and they must also be of the form that enables depth—organized complexity—to emerge. That the laws of our universe possess this crucial dual property is a fact of literally cosmic significance.

6

The Mathematical Secret

THE ASTRONOMER JAMES JEANS once proclaimed that God is a mathematician. His pithy phrase expresses in metaphorical terms an article of faith adopted by almost all scientists today. The belief that the underlying order of the world can be expressed in mathematical form lies at the very heart of science, and is rarely questioned. So deep does this belief run that a branch of science is considered not to be properly understood until it can be cast in the impersonal language of mathematics.

As we have seen, the idea that the physical world is the manifestation of mathematical order and harmony can be traced back to ancient Greece. It came of age in Renaissance Europe with the work of Galileo, Newton, Descartes, and their contemporaries. "The book of nature," opined Galileo, "is written in mathematical language." Why this should be so is one of the great mysteries of the universe. The physicist Eugene Wigner has written of the "unreasonable effectiveness of mathematics in the natural sciences," quoting C. S. Pierce that "it is probable that there is some secret here which remains to be discovered."[1] A recently published book[2] devoted to this topic containing essays by nineteen scientists (this author included) failed to uncover the secret, or even to arrive at any consensus. Opinions ranged from those who maintain that human beings have simply invented mathematics to fit the facts of experience, to those who are convinced that there is a deep and meaningful significance behind nature's mathematical face.

The Mathematical Secret

Is Mathematics Already "Out There"?

Before we tackle the topic of its "unreasonable effectiveness," it is important to have some understanding of what mathematics is. There are two broadly opposed schools of thought concerning its character. The first of these holds that mathematics is purely a human invention, the second that it has an independent existence. We have already met one version of the "invention," or formalist, interpretation in chapter 4, in the discussion of Hilbert's program for the mechanization of theorem-proving. Before the work of Gödel it was possible to believe that mathematics is an entirely formal exercise, consisting of nothing more than a vast collection of logical rules that link one set of symbols to another. This edifice was regarded as a completely self-contained structure. Any connection with the natural world was considered to be coincidental and of no relevance whatever to the mathematical enterprise itself, this being concerned only with the elaboration and exploration of the consequences of the formal rules. As explained in the previous chapters, Gödel's incompleteness theorem put paid to this strictly formalist position. Nevertheless, many mathematicians retain the belief that mathematics is only an invention of the human mind, having no meaning beyond that attributed to it by mathematicians.

The opposing school is known as Platonism. Plato, it will be recalled, had a dualistic vision of reality. On the one hand stood the physical world, created by the Demiurge, fleeting and impermanent. On the other stood the realm of Ideas, eternal and unchanging, acting as a sort of abstract template for the physical world. Mathematical objects he considered to belong to this Ideal realm. According to Platonists, we do not invent mathematics, we *discover* it. Mathematical objects and rules enjoy an independent existence: they transcend the physical reality that confronts our senses.

To sharpen the focus of this dichotomy, let us look at a specific example. Consider the statement "Twenty-three is the smallest prime number greater than twenty." The statement is either true or false. In fact, it is true. The question before us is whether the statement is true in a timeless, absolute sense. Was the statement true before the invention/discovery of prime numbers? The Platonist would answer yes, because prime numbers exist, abstractly, whether human beings know about them or not. The formalist would dismiss the question as meaningless.

What do professional mathematicians think? It is often said that mathematicians are Platonists on weekdays and formalists at weekends. While actually working on mathematics, it is hard to resist the impression that one is actually engaged in the process of discovery, much as in an experimental science. The mathematical objects take on a life of their own, and often display totally unexpected properties. On the other hand, the idea of a transcendent realm of mathematical Ideas seems too mystical for many mathematicians to admit, and if challenged they will usually claim that when engaging in mathematical research they are only playing games with symbols and rules.

Nevertheless, some distinguished mathematicians have been self-confessed Platonists. One of these was Kurt Gödel. As might be expected, Gödel based his philosophy of mathematics on his work on undecidability. He reasoned that there will always be mathematical statements that are true but can never be proved to be true from existing axioms. He envisaged these true statements as therefore already existing "out there" in a Platonic domain, beyond our ken. Another Platonist is the Oxford mathematician Roger Penrose. "Mathematical truth is something that goes beyond mere formalism," he writes.[3] "There often does appear to be some profound reality about these mathematical concepts, going quite beyond the deliberations of any particular mathematician. It is as though human thought is, instead, being guided towards some eternal external truth—a truth which has a reality of its own, and which is revealed only partially to any one of us." Taking as an example the system of complex numbers, Penrose feels that it has "a profound and timeless reality."[4]

Another example that has inspired Penrose to adopt Platonism is something called "the Mandelbrot set" after the IBM computer scientist Benoit Mandelbrot. The set is actually a geometrical form known as a "fractal," which is closely related to the theory of chaos, and provides another magnificent example of how a simple recursive operation can produce an object of fabulously rich diversity and complexity. The set is generated by successive applications of the rule (or mapping) $z \rightarrow z^2 + c$, where z is a complex number and c is a certain fixed complex number. The rule simply means: pick a complex number z and replace it with $z^2 + c$, then take this number to be z and make the same replacement, and so on, again and again. The successive complex numbers can be plotted on a sheet of paper (or a computer screen) as the rule is applied, each number represented as a dot. What is found is that for some choices of c the dot soon leaves the screen. For other choices, however, the dot wanders about forever within a bounded region. Now, each choice of c itself corresponds to a dot on the screen. The collection of all such c-dots forms the Mandelbrot set. This set has such an extraordinarily complicated structure that it is impossible to convey in words its awesome beauty. Many examples of portions of the set have been used for artistic displays. A distinctive feature of the Mandelbrot set is that any portion of it may be magnified again and again without limit, and each new layer of resolution brings forth new riches and delights.

Penrose remarks that, when Mandelbrot embarked on his study of the set, he had no real prior conception of the fantastic elaboration inherent in it:

> The complete details of the complication of the structure of Mandelbrot's set cannot really be fully comprehended by any one of us, nor can it be fully revealed by any computer. It would seem that this structure is not just part of our minds, but it has a reality of its own. . . . The computer is being used in essentially the same way that the experimental physicist uses a piece of experimental apparatus to explore the structure of the physical world. The Mandelbrot set is not an invention of the human mind: it was a discovery. Like Mount Everest, the Mandelbrot set is just there![5]

Mathematician and well-known popularizer Martin Gardner concurs with this conclusion: "Penrose finds it incomprehensible (as do I) that anyone could suppose that this exotic structure is not as much 'out there' as Mount Everest is, subject to exploration in the way a jungle is explored."[6]

"Is mathematics invention or discovery?" asks Penrose. Do mathematicians get so carried away with their inventions that they imbue them with a spurious reality? "Or are mathematicians really uncovering truths which are, in fact, already 'there'—truths whose existence is quite independent of the mathematicians' activities?" In proclaiming his adherence to the latter point of view, Penrose points out that in cases such as the Mandelbrot set "much more comes out of the structure than is put in in the first place. One may take the view that in such cases the mathematicians have stumbled upon 'works of God.' " Indeed, he sees an analogy in this respect between mathematics and inspired works of art: "It is a feeling not uncommon amongst artists, that in their greatest works they are revealing eternal truths which have some kind of prior etherial existence. . . . I cannot help feeling that, with mathematics, the case for believing in some kind of etherial, eternal existence . . . is a good deal stronger."[7]

It is easy to gain the impression that there exists a huge landscape of mathematical structures, and that mathematicians explore this peculiar but inspiring territory, perhaps aided by the guiding hand of experience or the signpost of recent discoveries. Along the way these mathematicians come across new forms and theorems that are already there. The mathematician Rudy Rucker thinks of mathematical objects as occupying a sort of mental space—which he calls the "Mindscape"—just as physical objects occupy a physical space. "A person who does mathematical research," he writes, "is an explorer of the Mindscape in much the same way that Armstrong, Livingstone, or Cousteau are explorers of the physical features of our Universe." Occasionally different explorers will pass over the same terrain and report independently on their findings. "Just as we all share the same Universe, we all share the same Mindscape," believes Rucker.[8] John Barrow also cites the phenomenon of independent discovery in mathematics as evidence for "some objective element" that is independent of the psyche of the investigator.

Penrose conjectures that the way mathematicians make discoveries and communicate mathematical results to each other offers evidence of a Platonic realm, or Mindscape:

> I imagine that whenever the mind perceives a mathematical idea it makes contact with Plato's world of mathematical concepts. . . . When one 'sees' a mathematical truth, one's consciousness breaks through into this world of ideas, and makes direct contact with it. . . . When mathematicians communicate, this is made possible by each one having *a direct route to truth,* the consciousness of each being in a position to perceive mathematical truths directly, through this process of 'seeing.' Since each can make contact with Plato's world directly, they can more readily communicate with each other than one might have expected. The mental images that each one has, when making this Platonic contact, might be rather different in each case, but communication is possible because each is directly in contact with the *same* eternally existing Platonic world![9]

Sometimes this "breaking-through" is sudden and dramatic, and affords what is usually referred to as mathematical inspiration. The French mathematician Jacques Hadamard made a study of this phenomenon, and cites the case of Carl Gauss, who had for years been wrestling with a problem about whole numbers: "Like a sudden flash of lightning, the riddle happened to be solved. I myself cannot say what was the conducting thread which connected what I previously knew with what made my success possible."[10] Hadamard also gives the famous case of Henri Poincaré, who had likewise spent a lot of time fruitlessly tackling a problem concerning certain mathematical functions. One day Poincaré set out on a geological excursion, and went to board the bus. "At the moment when I put my foot on the step, the idea came to me, without anything in my former thoughts seeming to have paved the way for it," he reported.[11] So certain was he that the problem was solved that he put it to the back of his mind and continued his conversation. When he returned from the trip he was able to prove the result readily, at his leisure.

Penrose recounts a similar incident about his work on black holes and space-time singularities.[12] He was engaged in conversation in a London street, and was about to cross a busy road when the crucial idea occurred to him, but only fleetingly, so that when he resumed the conversation on the other side of the road the idea was blotted out. It was only later that he became aware of a curious feeling of elation, and mentally recounted the events of the day. Finally, he remembered the brief inspirational flash, and knew it was the key to the problem that had been occupying his attention for a long while. It was only some time later that the correctness of the idea was rigorously demonstrated.

Many physicists share this Platonic vision of mathematics. For example, Heinrich Hertz, the first to produce and detect radio waves in the laboratory, once said: "One cannot escape the feeling that these mathematical formulas have an independent existence of their own, and they are wiser than even their discoverers, that we get more out of them than was originally put into them."[13]

I once asked Richard Feynman whether he thought of mathematics and, by extension, the laws of physics as having an independent existence. He replied:

The problem of existence is a very interesting and difficult one. If you do mathematics, which is simply working out the consequences of assumptions, you'll discover for instance a curious thing if you add the cubes of integers. One cubed is one, two cubed is two times two times two, that's eight, and three cubed is three times three times three, that's twenty-seven. If you add the cubes of these, one plus eight plus twenty-seven—let's stop here—that would be thirty-six. And that's the square of another number, six, and that number is the sum of those same integers, one plus two plus three. . . . Now, that fact which I've just told you about might not have been known to you before. You might say: "Where is it, what is it, where is it located, what kind of reality does it have?" And yet you came upon it. When you discover these things, you get the feeling that they were true before you found them. So you get the idea that somehow they existed somewhere, but there's nowhere for such things. It's just a feeling. . . . Well, in the case of physics we have double trouble. We come upon these mathematical interrelationships but they apply to the universe, so the problem of where they are is doubly confusing. . . . Those are philosophical questions that I don't know how to answer.[14]

The Cosmic Computer

In recent years, deliberations on the nature of mathematics have increasingly come under the influence of computer scientists, who have their own special view of the subject. Not surprisingly, perhaps, many computer scientists regard the computer as a central component in any system of thought that attempts to give meaning to mathematics. In its extreme form this philosophy proclaims, "What can't be computed is meaningless." In particular, any description of the physical universe must use mathematics that can actually be implemented, in principle, by a computer. Clearly this rules out theories of the sort described in chapter 5, involving predictions of uncomputable numbers for physical quantities. No mathematical operations that involve an infinite number of steps can be permitted. This rules out large areas of mathematics, much of which has been applied to physical systems. More seriously, even those mathematical results that imply a finite but very large number of steps are suspect, if one supposes that the computing power of the universe is limited. Rolf Landauer is an exponent of this viewpoint: "Not only does physics determine what computers can do, but what computers can do, in turn, will define the ultimate nature of the laws of physics. After all, the laws of physics are algorithms for the processing of information, and will be futile, unless these algorithms are implementable in our universe, with its laws and resources."[15]

If meaningful mathematics depends on the resources available to the universe, there are far-reaching implications. According to standard cosmological theory, light can have traveled only a finite distance since the origin of the universe (basically because the universe is of a finite age). But no physical object or influence, and in particular no information, can exceed the speed of light. It follows that the region of the universe to which we are causally attached contains only a finite number of particles. The outer limit to this region is known as our horizon. It is the most distant surface in space to which light emitted from our vicinity of the universe at the time of the big bang can have now reached. When it comes to computation, obviously only those regions of the universe between which information can flow can be considered as part of a single computing system; this will be the region within our horizon. Imagine that every particle in this region is commandeered and incorporated into a gigantic cosmic computer. Then even this awesome machine would still have limited computational capabilities, because it contains a finite number of particles (about 10^{80}, in fact). It could not, for example, even compute pi to infinite precision. According to Landauer, if the universe as a whole can't compute it, forget it. So "humble pi" would no longer be a precisely defined quantity. This carries the implication that the ratio of the circumference of a circle to its diameter could not be considered to be a precise, fixed number—even in the idealized case of perfect geometrical lines—but would be subject to uncertainty.

Stranger still is the fact that, because the horizon expands with time as light moves outward into space, the resources available to the region within the horizon would have been less in the past. This implies that mathematics is *time-dependent*, a notion diametrically opposed to Plato's view that mathematical truths are timeless, transcendent, and eternal. For example, at one second after the big bang, a horizon volume would have contained only a tiny fraction of the present number of atomic particles. At the so-called Planck time (10^{-43}) a horizon volume typically contained only one particle. The computing power of the universe at the Planck time would therefore have been essentially zero. Pursuing the Landauer philosophy to its logical conclusion, this suggests that all mathematics was meaningless at that epoch. If so, then attempts to apply mathematical physics to the early universe—in particular the whole program of quantum cosmology and the cosmic origin described in chapter 2—are also rendered meaningless.

Why Us?

"The only incomprehensible thing about the universe is that it is comprehensible."

Albert Einstein

The success of the scientific enterprise can often blind us to the astonishing fact that science works. Although most people take it for granted, it is both incredibly fortunate and incredibly mysterious that we are able to fathom the workings of nature by use of the scientific method. As I have already explained, the essence of science is to uncover patterns and regularities in nature by finding algorithmic compressions of observations. But the raw data of observation rarely exhibit explicit regularities. Instead we find that nature's order is hidden from us, it is written in code. To make progress in science we need to crack the cosmic code, to dig beneath the raw data and uncover the hidden order. I often liken fundamental science to doing a crossword puzzle. Experiment and observation provide us with clues, but the clues are cryptic, and require some considerable ingenuity to solve. With each new solution, we glimpse a bit more of the overall pattern of nature. As with a crossword, so with the physical universe, we find that the solutions to independent clues link together in a consistent and supportive way to form a coherent unity, so that, the more clues we solve, the easier we find it to fill in the missing features.

What is remarkable is that human beings are actually able to carry out this code-breaking operation, that the human mind has the necessary intellectual equipment for us to "unlock the secrets of nature" and make a passable attempt at completing nature's "cryptic crossword." It would be easy to imagine a world in which the regularities of nature were transparent and obvious to all at a glance. We can also imagine another world in which either there were no regularities, or the regularities were so well hidden, so subtle, that the cosmic code would require vastly more brainpower than humans possess. But instead we find a situation in which the difficulty of the cosmic code seems almost to be attuned to human capabilities. To be sure, we have a pretty tough struggle decoding nature, but so far we have had a good deal of success. The challenge is just hard enough to attract some of the best brains available, but not so hard as to defeat their combined efforts and deflect them onto easier tasks.

The mystery in all this is that human intellectual powers are presumably determined by biological evolution, and have absolutely no connection with doing science. Our brains have evolved in response to environmental pressures, such as the ability to hunt, avoid predators, dodge falling objects, etc. What has this got to do with discovering the laws of electromagnetism or the structure of the atom? John Barrow is also mystified: "Why should our cognitive processes have tuned themselves to such an extravagant quest as the understanding of the entire Universe?" he asks. "Why should it be *us*? None of the sophisticated ideas involved appear to offer any selective advantage to be exploited during the pre-conscious period of our evolution. . . . How fortuitous that our minds (or at least the minds of some) should be poised to fathom the depths of Nature's secrets."[16]

The mystery of our uncanny success in making scientific progress is deepened by the limitations of human educational development. On the one hand, there is a limit to the rate at which we can grasp new facts and concepts, especially those of an abstract quality. It generally requires at least fifteen years of studying for a student to achieve a sufficient grasp of mathematics and science to make a real contribution to fundamental research. Yet it is well known that, especially in mathematical physics, the major advances are made by men and women in their twenties, or at best early thirties. Newton, for example, was only twenty-four when he hit upon the law of gravitation. Dirac was still a Ph.D. student when he formulated his relativistic wave equation that led to the discovery of antimatter. Einstein was twenty-six when he put together the special theory of relativity, the foundations of statistical mechanics, and the photoelectric effect in a few glorious months of creative activity. Though older scientists are quick to deny it, there is strong evidence that truly innovative creativity in science fades away in the middle years. The combination of educational progress and waning creativity hem in the scientist, providing a brief but crucial "window of opportunity" in which to make a contribution. Yet these intellectual restrictions presumably have their roots in mundane aspects of evolutionary biology, connected with the human lifespan, the structure of the brain, and the social organization of our species. How odd, then, that the durations involved are such as to permit creative scientific endeavor.

Again, it is easy to imagine a world in which we all had plenty of time to learn the necessary facts and concepts to do fundamental science, or another world in which it would take so many years to learn all the necessary things that death would intervene, or one's creative years would have passed, long before the educational phase was finished. And no feature of this uncanny "tuning" of the human mind to the workings of nature is more striking than mathematics, the product of the human mind that is somehow linked into the secrets of the universe.

Why Are the Laws of Nature Mathematical?

Few scientists stop to wonder why the fundamental laws of the universe are mathematical; they just take it for granted. Yet the fact that "mathematics works" when applied to the physical world—and works so stunningly well—demands explanation, for it is not clear we have any absolute right to expect that the world should be well described by mathematics. Although most scientists assume the world must be that way, the history of science cautions against this. Many aspects of our world have been taken for granted, only to be revealed as the result of special conditions or circumstances. Newton's concept of absolute, universal time is a classic example. In daily life this picture of time serves us well, but it turns out to work well only because we move about much slower than light. Might mathematics work well because of some other special circumstances?

One approach to this conundrum is to regard the "unreasonable effectiveness" of mathematics—to use Wigner's phrase—as a purely cultural phenomenon, a result of the way in which human beings have chosen to think about the world. Kant cautioned that if we view the world through rose-tinted spectacles it's no surprise if the world looks rosy. We are prone, he maintained, to project onto the world our own mental bias toward mathematical concepts. In other words, we read mathematical order into nature rather than read it out of nature. This argument has some force. There is no doubt that scientists prefer to use mathematics when studying nature, and tend to select those problems that are amenable to a mathematical treatment. Those aspects of nature that are not readily captured by mathematics (e.g., biological and social systems) are apt to be de-emphasized. There is a tendency to describe as "fundamental" those features of the world that fall into this mathematizable category. The question "Why are the fundamental laws of nature mathematical?" then invites the trivial answer: "Because we define to be fundamental those laws that are mathematical."

Our view of the world will obviously be determined in part from the way our brains are structured. For reasons of biological selection we can scarcely guess at, our brains have evolved to recognize and focus on those aspects of nature that display mathematical patterns. As I remarked in chapter 1, it is possible to imagine alien life forms with a completely different evolutionary history, and brains that bear little resemblance to ours. These aliens might not share our categories of thought, including our love of mathematics, and would see the world in ways that would be utterly incomprehensible to us.

So is the success of mathematics in science just a cultural quirk, an accident of our evolutionary and social history? Some scientists and philosophers have claimed that it is, but I confess I find this claim altogether too glib, for a number of reasons. First, much of the mathematics that is so spectacularly effective in physical theory was worked out as an abstract exercise by pure mathematicians long before it was applied to the real world. The original investigations were entirely unconnected with their eventual application. This "independent world created out of pure intelligence," as James Jeans expressed it, was later found to have use in describing nature. The British mathematician G. H. Hardy wrote that he practiced mathematics for its beauty, not its practical value. He stated almost proudly that he could foresee no useful application whatever for his work. And yet we discover, often years afterward, that nature is playing by the very same mathematical rules that these pure mathematicians have already formulated. (This includes, ironically, much of Hardy's work too). Jeans pointed out that mathematics is only one of many systems of thought. There have been attempts to build models of the universe as a living organism, for example, or as a machine. These made little progress. Why should the mathematical approach prove so fruitful if it does not uncover some real property of nature?

Penrose has also considered this topic, and rejects the cultural viewpoint. Referring to the astonishing success of theories such as the general theory of relativity, he writes:

> It is hard for me to believe, as some have tried to maintain, that such SUPERB theories could have arisen merely by some random natural selection of ideas leaving only the good ones as survivors. The good ones are simply much too good to be the survivors of ideas that have arisen in that random way. There must, instead, be some deep underlying reason for the accord between mathematics and physics, i.e. between Plato's world and the physical world. [17]

Penrose endorses the belief, which I have found to be held by most scientists, that major advances in mathematical physics really do represent discoveries of some genuine aspect of reality, and not just the reorganization of data in a form more suitable for human intellectual digestion.

It has also been argued that the structure of our brains has evolved to reflect the properties of the physical world, including its mathematical content, so that it is no surprise that we discover mathematics in nature. As already remarked, it is certainly a surprise, and a deep mystery, that the human brain has evolved its extraordinary mathematical ability. It is very hard to see how abstract mathematics has any survival value. Similar comments apply to musical ability.

We come to know about the world in two quite distinct ways. The first is by direct perception, the second by application of rational reasoning and higher intellectual functions. Consider observing the fall of a stone. The physical phenomenon taking place in the external world is mirrored in our minds because our brains construct an internal mental model of the world in which an entity corresponding to the physical object "stone" is perceived to move through three-dimensional space: we *see* the stone fall. On the other hand, one can know about the fall of the stone in an entirely different and altogether more profound way. From a knowledge of Newton's laws plus some appropriate mathematics one could produce another sort of model of the fall of the stone. This is not a mental model in the sense of perception; nevertheless it is still a mental construct, and one which links the specific phenomenon of the fall of the stone to a wider body of physical processes. The mathematical model using the laws of physics is not something we actually *see,* but it is, in its own abstract way, a type of knowledge of the world, and, moreover, knowledge of a higher order.

It seems to me that Darwinian evolution has equipped us to know the world by direct perception. There are clear evolutionary advantages in this, but there is no obvious connection at all between this sort of sensorial knowledge and intellectual knowledge. Students often struggle with certain branches of physics, like quantum mechanics and relativity, because they try to understand these topics by mental visualization. They attempt to "see" curved space or the activity of an atomic electron in the mind's eye, and fail completely. This is not due to inexperience—I don't believe any human being can really form an accurate visual image of these things. Nor is this a surprise—quantum and relativity physics are not especially relevant to daily life, and there is no selective advantage in our having brains able to incorporate quantum and relativistic systems in our mental model of the world. In spite of this, however, physicists are able to reach an understanding of the worlds of quantum physics and relativity by the use of mathematics, selected experimentation, abstract reasoning, and other rational procedures. The mystery is, why do we have this dual capability for knowing the world? There is no reason to believe that the second method springs from a refinement of the first. They are entirely independent ways of coming to know about things. The first serves an obvious biological need, the latter is of no apparent biological significance at all.

The mystery becomes even deeper when we take account of the existence of mathematical and musical geniuses, whose prowess in these fields is orders of magnitude better than that of the rest of the population. The astonishing insight of mathematicians such as Gauss and Riemann is attested not only by their remarkable mathematical feats (Gauss was a child prodigy and also had a photographic memory), but also by their ability to write down theorems without proof, leaving later generations of mathematicians to struggle over the demonstrations. How these mathematicians were able to come up with their results "ready-made," when the proofs often turned out to involve volumes of complex mathematical reasoning, is a major puzzle.

Probably the most famous case is that of the Indian mathematician S. Ramanujan. Born in India in the late nineteenth century, Ramanujan came from a poor family and had only a limited education. He more or less taught himself mathematics and, being isolated from mainstream academic life, he approached the subject in a very unconventional manner. Ramanujan wrote down a great many theorems without proof, some of them of a very peculiar nature that would not normally have occurred to more conventional mathematicians. Eventually some of Ramanujan's results came to the attention of Hardy, who was astonished. "I have never seen anything in the least like them before," he commented. "A single look at them is enough to show that they could only be written down by a mathematician of the highest class." Hardy was able to prove some of Ramanujan's theorems by deploying the full range of his own considerable mathematical skills, but only with the greatest difficulty. Other results defeated him completely. Nevertheless, he felt they must be correct, for "no one would have the imagination to invent them." Hardy subsequently arranged for Ramanujan to travel to Cambridge to work with him. Ramanujan unfortunately suffered from culture shock and medical problems, and he died prematurely at the age of only thirty-three, leaving a vast stock of mathematical conjectures for posterity. To this day nobody really knows how he achieved his extraordinary feats. One mathematician commented that the results just seemed "to flow from his brain" effortlessly. This would be remarkable enough in any mathematician, but in one who was largely unfamiliar with conventional mathematics it is truly extraordinary. It is very tempting to suppose that Ramanujan had a particular faculty that enabled him to view the mathematical Mindscape directly and vividly, and pluck out ready-made results at will.

Scarcely less mysterious are the weird cases of so-called lightning calculators—people who can perform fantastic feats of mental arithmetic almost instantly, without the slightest idea of how they arrive at the answer. Shakuntala Devi lives in Bangalore in India but regularly travels the world, amazing audiences with feats of mental arithmetic. On one memorable occasion in Texas she correctly found the twenty-third root of a two-hundred-digit number in fifty seconds!

Even more peculiar, perhaps, are the cases of "autistic savants," people who are mentally handicapped and may have difficulty performing even the most elementary formal arithmetic manipulations, but who nevertheless possess the uncanny ability to produce correct answers to mathematical problems that appear to ordinary people to be impossibly hard. Two American brothers, for instance, can consistently outdo a computer in finding prime numbers even though they are both mentally retarded. In another case, featured on British television, a handicapped man correctly and almost instantly gave the day of the week when presented with any date, even from another century!

We are, of course, used to the fact that all human abilities, physical and mental, show wide variations. Some people can jump six feet off the ground, whereas most of us can manage barely three. But imagine someone coming along and jumping sixty feet, or six hundred feet! Yet the intellectual leap represented by mathematical genius is far in excess of these physical differences.

Scientists are a long way from understanding how mental abilities are controlled by our genes. Perhaps only very rarely do humans contain the genetic imprint that codes for fantastic mathematical powers. Or perhaps it is not so rare, but the relevant genes are usually not switched on. However, whatever is the case, the necessary genes are present in the human gene pool. The fact that mathematical geniuses occur in every generation suggests that this quality is a rather stable factor in the gene pool. If this factor has evolved by accident rather than in response to environmental pressure, then it is a truly astonishing coincidence that mathematics finds such ready application to the physical universe. If, on the other hand, mathematical ability does have some obscure survival value and has evolved by natural selection, we are still faced with the mystery of why the laws of nature are mathematical. After all, surviving "in the jungle" doesn't require knowledge of the *laws* of nature, only of their manifestations. We have seen how the laws themselves are in code, and not connected in a simple way at all to the actual physical phenomena subject to those laws. Survival depends on an appreciation of how the world is, not of any hidden underlying order. Certainly it cannot depend on the hidden order within atomic nuclei, or in black holes, or in subatomic particles that are produced on Earth only inside particle-accelerator machines.

It might be supposed that when we duck to avoid a missile, or judge how fast to run to jump a stream, we are making use of a knowledge of the laws of mechanics, but this is quite wrong. What we use are previous experiences with similar situations. Our brains respond automatically when presented with such challenges; they don't integrate the Newtonian equations of motion in the way the physicist does when analyzing these situations scientifically. To make judgments about motion in three-dimensional space, the brain needs certain special properties. To do mathematics (such as the calculus needed to describe this motion) also requires special properties. I see no evidence for the claim that these two apparently very different sets of properties are actually the same, or that one follows as a (possibly accidental) byproduct of the other.

In fact, all the evidence is to the contrary. Most animals share our ability to avoid missiles and jump effectively, yet they display no significant mathematical ability. Birds, for example, are far more adept at exploiting the laws of mechanics than humans, and their brains have evolved very sophisticated qualities as a result. But experiments with birds' eggs have demonstrated that birds cannot even count to more than about three. Awareness of the regularities of nature, such as those manifested in mechanics, has good survival value, and is wired into animal and human brains at a very primitive level. By contrast, mathematics as such is a higher mental function, apparently unique to humans (as far as terrestrial life is concerned). It is a product of the most complex system known in nature. And yet the mathematics it produces finds its most spectacularly successful applications in the most basic processes in nature, processes that occur at the subatomic level. Why should the most complex system be linked in this way to the most primitive processes of nature?

It might be argued that, as the brain is a product of physical processes, it should reflect the nature of those processes, including their mathematical character. But there is, in fact, no direct connection between the laws of physics and the structure of the brain. The thing which distinguishes the brain from a kilogram of ordinary matter is its complex organized form, in particular the elaborate interconnections between neurons. This wiring pattern cannot be explained by the laws of physics alone. It depends on many other factors, including a host of chance events that must have occurred during evolutionary history. Whatever laws may have helped shape the structure of the human brain (such as Mendel's laws of genetics), they bear no simple relationship to the laws of physics.

How Can We Know Something without Knowing Everything?

This question, posed many years ago by mathematician Hermann Bondi, is today even more problematical in the light of progress made in quantum theory. It is often said that nature is a unity, that the world is an interconnected whole. In one sense this is true. But it is also the case that we can frame a very detailed understanding of individual parts of the world without needing to know everything. Indeed, science would not be possible at all if we couldn't proceed in bite-sized stages. Thus the law of falling bodies discovered by Galileo did not require a knowledge of the distribution of all the masses in the universe; the properties of atomic electrons can be discovered without our needing to know the laws of nuclear physics. And so on. It is easy to imagine a world in which phenomena occurring at one location in the universe, or on one scale of size or energy, were intimately entangled with all the rest in a way that would forbid resolution into simple sets of laws. Or, to use the crossword analogy, instead of dealing with a connected mesh of separately identifiable words, we would have a single extremely complicated word answer. Our knowledge of the universe would then be an "all-or-nothing" affair.

The mystery is all the deeper for the fact that the separability of nature is actually only approximate. The universe *is*, in reality, an interconnected whole. The fall of an apple on Earth is affected by, and in turn reacts upon, the position of the moon. Atomic electrons are subject to nuclear influences. In both cases, however, the effects are tiny, and can be ignored for most practical purposes. But not all systems are like this. As I have explained, some systems are chaotic, and are exquisitely sensitive to the most minute external disturbances. It is this property that makes chaotic systems unpredictable. Yet, even though we live in a universe replete with chaotic systems, we are able to filter out a vast range of physical processes that are predictable and mathematically tractable.

The reason for this can be traced in part to two curious properties, called "linearity" and "locality." A linear system obeys certain very special mathematical rules of addition and multiplication associated with straight-line graphs—hence the word "linear"—which need not be spelled out here (see *The Matter Myth* for a detailed discussion). The laws of electromagnetism, which describe electric and magnetic fields and the behavior of light and other electromagnetic waves, are linear to a very high degree of approximation, for example. Linear systems cannot be chaotic, and are not highly sensitive to small external disturbances.

No system is *exactly* linear, so the issue of the separability of the world boils down to why nonlinear effects are in practice often so small. This is usually because the nonlinear forces concerned are either intrinsically very weak, or of very short range, or both. We do not know why the strengths and ranges of the various forces of nature are what they are. One day we may be able to compute them from some underlying fundamental theory. Alternatively, they might simply be "constants of nature" that cannot be derived from the laws themselves. A third possibility is that these "constants" are not God-given fixed numbers at all, but are determined by the actual state of the universe; in other words, they may be related to the cosmic initial conditions.

The property of locality has to do with the fact that, in most cases, the behavior of a physical system is determined entirely by the forces and influences that arise in its immediate vicinity. Thus, when an apple falls, its rate of acceleration at each point in space depends on the gravitational field only at that point. Similar remarks apply to most other forces and circumstances. There are, however, situations where nonlocal effects arise. In quantum mechanics, two subatomic particles can interact locally and then move very far apart. But the rules of quantum physics are such that, even if the particles end up on opposite sides of the universe, they must still be treated as an indivisible whole. That is, measurements performed on one of the particles will depend in part on the state of the other. Einstein referred to this nonlocality as "ghostly action-at-a-distance" and refused to believe it. But recent experiments have confirmed beyond doubt that such nonlocal effects are real. Generally speaking, at the subatomic level, where quantum physics is important, a collection of particles must be treated holistically. The behavior of one particle is inextricably entangled with those of the others, however great the interparticle separations may be.

This fact has an important implication for the universe as a whole. If one were to pick an arbitrary quantum state for the entire cosmos, it is probable that this state would represent a gigantic entanglement of all the particles in the universe. In chapter 2 I discussed the recent ideas of Hartle and Hawking concerning the quantum description of the whole universe—quantum cosmology. One of the great challenges to quantum cosmologists is to explain how the familiar world of experience has emerged from the fuzziness of its quantum origin. Quantum mechanics, it will be recalled, incorporates Heisenberg's uncertainty principle, which has the effect of smearing out the values of all observable quantities in an unpredictable way. Thus an electron in orbit around an atom cannot be considered to have a well-defined position in space at every moment. One should not really think of it as circling the atomic nucleus along a definite path, but instead as smeared out in an indeterminate manner around the nucleus.

Although this is the case for electrons in atoms, when it comes to macroscopic objects we do not observe such smearing. Thus the planet Mars has a definite position in space at each moment, and follows a definite orbit around the sun. In spite of this, Mars is still subject to the laws of quantum mechanics. One can now ask, as Enrico Fermi once did, why Mars is not smeared out around the sun in the same way as an electron is smeared out around an atom. In other words, given that the universe was born in a quantum event, how has an essentially non-quantum world emerged? When the universe originated, and was very small, quantum uncertainty engulfed it. Today, we do not notice any residual uncertainty in macroscopic bodies.

Most scientists have tacitly assumed that an approximately non-quantum (or "classical," to use the jargon) world would have emerged automatically from the big bang, even from a big bang in which quantum effects dominated. Recently, however, Hartle and Gell-Mann have challenged this assumption. They argue that the existence of an approximately classical world, in which well-defined material objects exist at distinct locations in space, and in which there is a well-defined concept of time, requires special cosmic initial conditions. Their calculations indicate that, for the majority of initial states, a generally classical world would *not* emerge. In that case the separability of the world into distinct objects occupying definite positions in a well-defined background space-time would not be possible. There would be no locality. It seems likely that in such a smeared-out world one could know nothing without knowing everything. Indeed, Hartle and Gell-Mann argue that the very notion of traditional laws of physics, such as Newtonian mechanics, should be regarded not as truly fundamental aspects of reality, but as *relics* of the big bang, and a consequence of the special quantum state in which the universe originated.

If it is also the case, as remarked briefly above, that the strengths and ranges of the forces of nature are likewise dependent on the quantum state of the universe, then we reach a remarkable conclusion. Both the linearity and the locality of most physical systems would not be a consequence of some fundamental set of laws at all, but would be due to the peculiar quantum state in which the universe originated. The intelligibility of the world, the fact that we can progressively discover laws and extend our understanding of nature—the very fact that science works—would not be an inevitable and absolute right, but could be traced to special, perhaps highly special, cosmic initial conditions. The "unreasonable effectiveness" of mathematics in its application to the natural world would then be due to unreasonably effective initial conditions.

7

Why Is the World
the Way It Is?

EINSTEIN ONCE REMARKED that the thing which most interested him was
whether God had any choice in creating the world as it is. Einstein was
not religious in the conventional sense, but he liked to use God as a
metaphor for expressing deep questions of existence. This particular
question has vexed generations of scientists, philosophers, and theologians. Does the world have to be the way it is, or could it have been
otherwise? And if it could have been otherwise, what sort of explanation should we seek for why it is as it is?

In referring to the question of God's freedom to create a world of his
choice, Einstein was alluding to the seventeenth-century philosopher
Benedict Spinoza. Spinoza was a pantheist, who regarded objects in the
physical universe as attributes of God rather than as God's creation. By
identifying God with nature, Spinoza rejected the Christian idea of a
transcendent Deity who created the universe as a free act. On the other
hand, Spinoza was no atheist: he believed he had a logical proof that
God *must* exist. Because he identified God with the physical universe,
this amounted to a proof that our particular universe must also exist.
For Spinoza, God had no choice in the matter: "Things could not have
been brought into being by God in any manner or in any order different
from that which has in fact obtained," he wrote.

This type of thinking—that things are as they are as a result of some
sort of logical necessity or inevitability—is quite common today among
scientists. Mostly, though, they prefer to drop God out of it altogether.
If they are right, it implies that the world forms a closed and complete

system of explanation, in which everything is accounted for and no mystery remains. It also means that in principle we need not actually observe the world to be able to work out its form and content: because everything follows from logical necessity, the nature of the universe would be deducible from reason alone. "I hold it true," wrote Einstein when flirting with this idea, "that pure thought can grasp reality, as the Ancients dreamed. . . . We can discover by means of purely mathematical constructions the concepts and the laws connecting them with each other, which furnish the key to the understanding of natural phenomena."[1] Of course, we may never be clever enough actually to derive the correct concepts and laws from mathematical deduction alone, but that is not the point. If such a closed explanatory scheme were even possible, it would profoundly alter our thinking about the universe and our place within it. But do these claims of completeness and uniqueness have any foundation, or are they just a vague hope?

An Intelligible Universe

Underlying all these questions is a crucial assumption: that the world is both rational and intelligible. This is often expressed as the "principle of sufficient reason," which states that everything in the world is as it is for some reason. Why is the sky blue? Why do apples fall? Why are there nine planets in the solar system? We are not usually satisfied with the reply: "Because that's just the way it is." We believe that there must be some reason why it is like that. If there are facts about the world that must simply be accepted without reason (so-called brute facts), then rationality breaks down and the world is absurd.

Most people accept the principle of sufficient reason without question. The entire scientific enterprise, for example, is built upon the assumed rationality of nature. Most theologians also adhere to the principle, because they believe in a rational God. But can we be absolutely sure that the principle is infallible? Is there sufficient reason to believe the principle of sufficient reason? To be sure, it usually works all right: apples fall because of gravity, the sky is blue because short-wavelength light is scattered by air molecules, and so on. But that does not guarantee it will always work. Of course, if the principle is false, then further inquiry into ultimate issues becomes pointless. Anyway, whether or not the principle is infallible, it is worth accepting it as a working hypothesis to see where it leads.

In confronting the deep issues of existence, we have to consider the possibility of two distinct classes of things.

In the first class are facts about the physical universe, such as the number of planets in the solar system. It is a matter of fact that there are nine planets, but it seems unreasonable to suppose that there *have* to be nine. Certainly we can easily imagine there being eight, or ten. A typical explanation for why there are nine might focus on the way in which the solar system formed from a cloud of gas, the relative abundance of the elements in the gas, and so on. Because an explanation for the features of the solar system depends on something other than itself, these features are said to be "contingent." Something is contingent if it could have been otherwise, so that the reason why it is the way that it is depends upon something else, something beyond itself.

The second class refers to facts or objects or events that are not contingent. Such things are called "necessary." Something is necessary if it is what it is quite independently of anything else. A necessary thing contains the reason for itself within itself, and it would be completely unchanged if everything else were different.

It is difficult to be convinced that there are any necessary things in nature. Certainly all the physical objects we encounter in the world, and the events that befall them, depend in some way on the rest of the world, and so must be considered contingent. Furthermore, if something is necessarily what it is, then it must always be what it is: it cannot change. A necessary thing can make no reference to time. Yet the state of the world continually changes with time, so all physical things that partake of that change must be contingent.

What about the universe as a whole, if we include in the definition of "universe" time itself? Might that be necessary? This is what Spinoza and his followers have claimed. At first glance it is hard to see that they can be right. We can easily imagine the universe to be different from what it is. Of course, simply being able to imagine something is no guarantee that such a thing is possible, even logically possible. But I believe that there are very good reasons why the universe might have been otherwise, as I shall shortly discuss.

What about the laws of physics? Are they necessary or contingent? Here the situation is less clear. Normally these laws are considered to be timeless and eternal, so perhaps a case could be made that they are necessary. On the other hand, experience shows that, as physics progresses, so what were thought to be independent laws are found to be linked together. A good example is the recent discovery that the weak nuclear force and the electromagnetic force are actually two aspects of a single electroweak force described by a common system of equations. So the individual forces turn out to be contingent on other forces. But is it possible that there exists a superforce, or even a completely unifying superlaw, that is necessary? Many physicists think so. Some contemporary scientists, such as the Oxford chemist Peter Atkins, point to this convergence of fundamental physics toward a unifying superlaw to argue that the physical world is not contingent but necessarily as it is. They maintain that it is gratuitous to seek for further explanation in metaphysics. These scientists look forward to a time when all the laws of physics will be combined into a single mathematical scheme, and claim that this scheme will be the only logically self-consistent one available.

But others have directed attention to this same progressive unification and drawn the opposite conclusion. For example, Pope John Paul II has been deeply impressed by the spectacular progress made in linking the various elementary particles of matter and the four fundamental forces of nature, and recently saw fit to address a scientific conference on the wide implications:

> Physicists possess a detailed though incomplete and provisional knowledge of elementary particles and of the fundamental forces through which they interact at low and intermediate energies. They now have an acceptable theory unifying the electromagnetic and weak nuclear forces, along with much less adequate but still promising grand unified field theories which attempt to incorporate the strong nuclear interaction as well. Further in the line of this same development, there are already several detailed suggestions for the final stage, superunification, that is, the unification of all four fundamental forces, including gravity. Is it not important for us to note that in a world of such detailed specialization as contemporary physics there exists this drive towards convergence?[2]

The point about this convergence is the way it progressively hems in the acceptable laws of physics. Each new linkage that is fashioned demands a mutual interdependence and consistency between the laws governing the hitherto independent parts. The requirement that all theories are consistent with quantum mechanics and the theory of relativity, for example, already imposes strong constraints on the mathematical form that the laws can assume. This prompts the spec- ulation that one day, perhaps soon, the convergence may be com- pleted, and a totally unified account of all the laws of nature acquired. This is the idea of a so-called Theory of Everything, briefly mentioned in chapter 1.

A Unique Theory of Everything?

Is a Theory of Everything feasible? Many scientists think so. Indeed, some of them even believe we may almost have such a theory. They cite the currently popular superstring theory as a serious attempt to amal- gamate all the fundamental forces and particles of physics, as well as the structure of space and time, into a single, all-embracing mathematical scheme. In fact, this confidence is not new. There is a long history of attempts to construct completely unifying accounts of the world. In his book *Theories of Everything: The Quest for Ultimate Explanation,* John Barrow attributes the lure of such a theory to the passionate belief in a rational cosmos: that there is a graspable logic behind physical existence that can be compressed into a compelling and succinct form.

The question then arises of whether, in achieving this total unifi- cation, the theory becomes so tightly constrained by the requirements of mathematical consistency that it is rendered unique. If that were so, there could be only one unified system of physics, with its various laws fixed by logical necessity. The world, it is said, would be explained: Newton's laws, Maxwell's electromagnetic-field equations, Einstein's gravitational-field equations, and all the rest would follow inexorably from the requirements of logical consistency as surely as Pythagoras' theorem follows from the axioms of Euclidean geometry. Taking this line of argument to its extreme, scientists needn't bother with obser- vation or experiment. Science would no longer be an empirical matter, but a branch of deductive logic, with the laws of nature acquiring the status of mathematical theorems, and the properties of the world deducible by the application of reason alone.

The belief that the nature of things in the world can be known solely through the exercise of pure reason, by using a deductive logical argument from self-evident premises, has a long history. Elements of such an approach can be found in the writings of Plato and Aristotle. It resurfaced in the seventeenth century with the rationalist philosophers, such as Descartes, who constructed a system of physics which he intended to be rooted in reason alone, rather than empirical observation. Much later, in the 1930s, the physicist E. A. Milne likewise attempted to construct a deductive description of gravitation and cosmology. In recent years, the idea that a completely unified description of physics might turn out to be deductively provable has once more become fashionable, and it was this that prompted Stephen Hawking to choose for his inaugural lecture for the Lucasian Chair the provocative title "Is the End in Sight for Theoretical Physics?"

But what evidence is there that such a state of affairs is likely? Leaving aside the uncertainties over whether the recent work on superstrings and the like really do point to an early unification, I believe it is demonstrably wrong that a superunified theory would be unique. I have arrived at this conclusion for a number of reasons. The first of these is that theoretical physicists frequently discuss mathematically consistent "toy universes" which certainly do not correspond to our universe. I explained the reason for this in chapter 1. We have already encountered one such toy universe—the cellular automaton. There are many others. It seems to me that, to have any hope of uniqueness, one would need to demand not just self-consistency, but a host of contingent specifications, such as conformity with relativity, or the presence of certain symmetries, or the existence of three dimensions of space and one of time.

The second problem concerns the very notion of logical and mathematical uniqueness. Mathematics must be founded upon a set of axioms. Though the theorems of mathematics may be deduced from within the system of axioms, the axioms themselves cannot. They must be justified from outside the system. One can imagine many different sets of axioms leading to different logical schemes. There is also the serious problem of Gödel's theorem. Recall that, according to this theorem, it is generally impossible, from within the system of axioms, even to prove that the axioms are consistent. And if consistency can be shown, then the system of axioms would not be complete, in the sense that there would exist true mathematical statements that could not be proved to be true within that system. In a recent article, Russell Stannard discussed the implications for the unification of physics:

A genuine theory of everything must explain not only how our universe came into being, but also why it is the only type of universe that there could have been—why there could only be one set of physical laws.

This goal I believe to be illusory. . . . This inherent, unavoidable lack of completeness must reflect itself in whatever mathematical system models our universe. As creatures belonging to the physical world, we will be included as part of that model. It follows that we shall never be able to justify the choice of axioms in the model—and consequently the physical laws to which those axioms correspond. Nor shall we be able to account for all true statements that can be made about the universe.[3]

John Barrow also examines the limitations that Gödel's theorem implies for a Theory of Everything, and concludes that such a theory would be "far from sufficient to unravel the subtleties of a Universe like ours. . . . There is no formula that can deliver all truth, all harmony, all simplicity. No Theory of Everything can ever provide total insight. For, to see through everything would leave us seeing nothing at all."[4]

So the search for a genuinely unique Theory of Everything that would eliminate all contingency and demonstrate that the physical world must necessarily be as it is, seems to be doomed to failure on grounds of logical consistency. No rational system can be proved both consistent and complete. There will always remain some openness, some element of mystery, something unexplained. The philosopher Thomas Torrance chides those who fall for the temptation to believe that the universe is "some sort of perpetuum mobile, a self-existing, self-supporting, self-explaining magnitude, wholly consistent and complete in itself and thus imprisoned within a pointless circularity of inescapable necessities." He cautions that "there is no intrinsic reason in the universe why it should exist at all, or why it should be what it actually is: hence we deceive ourselves if in our natural science we think we can establish that the universe could only be what it is."[5]

Is it possible that the laws of our universe, while not logically unique, are nevertheless the only possible laws that can also give rise to complexity? Perhaps our universe is the only possible one in which biology is permitted, and hence in which conscious organisms could arise. This would then be the only possible *cognizable* universe. Or, to return to Einstein's question about whether God had any choice in his creation, the answer would be no, unless he wanted it to go unnoticed. This possibility is mentioned by Stephen Hawking in his book *A Brief History of Time:* "There may be only one, or a small number, of complete unified theories, such as the heterotic string theory, that are self-consistent and allow structures as complicated as human beings who can investigate the laws of nature and ask about the nature of God."[6]

It may be that there is no logical impediment to this weaker proposal; I don't know. But I do know that there is absolutely no evidence in favor of it. A case could perhaps be made that we live in the *simplest possible* cognizable universe—that is, the laws of physics are the *simplest* logically self-consistent set that permit self-reproducing systems. But even this watered-down goal seems unattainable. As we have seen in chapter 4, there are cellular-automaton worlds in which self-reproduction can occur, and the defining rules of these worlds are so simple it is hard to imagine that the ultimate unified laws of physics would be simpler.

Let me now turn to a more serious problem with the "unique-universe" argument, and one which is often glossed over. Even if the laws of physics were unique, it doesn't follow that the physical universe itself is unique. As explained in chapter 2, the laws of physics must be augmented by cosmic initial conditions. One possible set of initial conditions is the proposal by Hartle and Hawking discussed at the end of that chapter. Now, though this may be a natural choice, it is only one of an infinite range of possible choices. There is nothing in present ideas about "laws of initial conditions" remotely to suggest that their consistency with the laws of physics would imply uniqueness. Far from it. Hartle himself has argued that there are deep reasons of principle why there cannot be unique laws: "We construct our theories as part of the universe, not outside it, and this fact must inevitably limit the theories we construct. A theory of initial conditions, for example, must be simple enough that it can be stored within the universe." In conducting our science, we move matter around. Even the process of thinking involves the disturbance of electrons in our brains. These disturbances, though minute, nevertheless affect the fate of other electrons and atoms in the universe. Hartle concludes: "In view of this there must be many theories of initial conditions rendered indistinguishable by our act of constructing them."[7]

Another fly in the ointment concerns the fundamentally quantum nature of the world, with its inherent indeterminism. All candidate

Theories of Everything must incorporate this principle, which implies that the best any such theory could do would be to fix some sort of most likely world. The actual world would differ in a myriad of unpredictable ways on the subatomic scale. This could be a major factor even on a macroscopic scale. A single subatomic encounter, for example, can produce a biological mutation that may alter the course of evolution.

Contingent Order

It seems, then, that the physical universe does not have to be the way it is: it could have been otherwise. Ultimately, it is the assumption that the universe is *both* contingent *and* intelligible that provides the motivation for empirical science. For without the contingency we would in principle be able to explain the universe using logical deduction alone, without ever observing it. And without the intelligibility there could be no science. "It is the combination of contingency and intelligibility," writes the philosopher Ian Barbour, "which prompts us to search for new and unexpected forms of rational order."[8] Barbour points out that the contingency of the world is fourfold. First, the laws of physics themselves appear to be contingent. Second, the cosmological initial conditions could have been otherwise. Third, we know from quantum mechanics that "God plays dice"—i.e., there is a fundamental statistical element in nature. Finally, there is the fact that the universe exists. After all, however comprehensive our theories of the universe may be, there is no obligation for the world actually to instantiate that theory. This last point has been vividly expressed by Stephen Hawking: "Why does the universe go to all the bother of existing?" he asks. "What is it that breathes fire into the equations and makes a universe for them to describe?"[9]

I believe that there is also a fifth type of contingency, which is to be found in the "higher-level" laws associated with the organizational properties of complex systems. I have given a complete account of what I mean by such laws in my book *The Cosmic Blueprint*, so I shall limit myself to a few examples. I have already mentioned Mendel's laws of genetics, which, though perfectly consistent with the underlying laws of physics, could not be derived solely from the laws of physics. Likewise, the various laws and regularities found in chaotic systems, or in self-organizing systems, depend not only on the laws of physics, but also on the specific nature of the systems concerned. In many cases the precise form of the patterns of behavior adopted by these systems depends upon some accidental microscopic fluctuation, and must therefore be considered as undetermined in advance. These higher-level laws and regularities thus possess important contingent features over and above the usual laws of physics.

The great mystery about contingency is not so much that the world could have been otherwise, but that it is contingently *ordered*. This is most forcefully apparent in the biological realm, where terrestrial organisms are clearly contingent in their particular forms (they could so easily have been different), yet where there is a conspicuous and pervasive order in the biosphere. If objects and events in the world were merely haphazard and arranged in no especially significant way, their *particular* arrangement would still be mysterious. But the fact that the contingent features of the world are also ordered or patterned is surely deeply meaningful.

Another highly relevant feature of the world's ordered contingency concerns the nature of that order, which is such as to bestow a rational unity on the cosmos. Moreover, this holistic orderliness is *intelligible* to us. These features make the mystery much, much deeper. But, whatever its explanation, the entire scientific enterprise is founded upon it. "It is this combination of contingence, rationality, freedom and stability of the universe," writes Torrance, "which gives it its remarkable character, and which makes scientific exploration of the universe not only possible for us but incumbent upon us. . . . It is through relying on the indissoluble bond between contingence and order in the universe that natural science has come to operate with the distinctive interconnection between experiment and theory which has characterized our greatest advances in knowledge of the physical world."[10]

My conclusion, then, is that the physical universe is not compelled to exist as it is; it could have been otherwise. In that case we are returned to the problem of why it is as it is. What sort of explanation might we seek for its existence and its remarkable form?

Let me first dispose of a rather trivial attempt at explanation that is sometimes proposed. It has been argued by some that everything in the universe can be explained in terms of something else, and that in terms of something else again and so on, in an infinite chain. As I mentioned in chapter 2, some proponents of the steady-state theory have used this reasoning, on the basis that the universe has no origin in time in this theory. However, it is quite wrong to suppose that an infinite chain of explanation is satisfactory on the basis that every member of that chain is explained by the next member. One is still left with the mystery of why that *particular* chain is the one that exists, or why any chain exists at all. Leibniz made this point eloquently by inviting us to consider an infinite collection of books, each one copied from a previous one. To say that the content of the book is thereby explained is absurd. We are still justified in asking who the author was.

It seems to me that, if one perseveres with the principle of sufficient reason and demands a rational explanation for nature, then we have no choice but to seek that explanation in something beyond or outside the physical world—in something metaphysical—because, as we have seen, a contingent physical universe cannot contain within itself an explanation for itself. What sort of metaphysical agency might be able to create a universe? It is important to guard against the naïve image of a Creator producing a universe at some instant in time by supernatural means, like a conjurer pulling a rabbit out of a hat. As I have explained at length, creation cannot consist of merely causing the big bang. We are searching instead for a more subtle, timeless notion of creation which, to use Hawking's phrase, breathes fire into the equations, and thus promotes the merely possible to the actually existing. This agency is creative in the sense of being somehow responsible for the laws of physics, which govern, among other things, how space-time evolves.

Naturally the theologians argue that the creative agency that provides an explanation for the universe is God. But what sort of agency would such a being be? If God were a mind (or Mind), we might fairly describe him as a person. But not all theists accept the need for this. Some prefer to think of God as Being-Itself or as a Creative Force, rather than as a Mind. Indeed, it may be that minds and forces are not the only agencies that have creative potency. The philosopher John Leslie has argued that "ethical requirement" might do the job, an idea that he traces back to Plato. In other words, the universe exists because it is good that it does so. "Belief in God," writes Leslie, "becomes belief that the universe exists because it ought to."[11] The idea seems strange. How can "ethical requirement" create a universe? Let me repeat, though, that we are not talking about creation in the causal, mechanical sense here, as when a builder builds a house. We are talking about "breathing fire" into the equations that encode the laws of physics, promoting the merely possible to the actual. What sort of entities can "breathe fire" in this sense? Clearly no familiar material thing. If there is to be an answer at all, it would have to be something pretty abstract and unfamiliar. There is no *logical* contradiction in attributing creative potency to ethical or aesthetic qualities. But neither is there any logical necessity to do so. Leslie proposes, however, that there may be a weaker, nonlogical, sense of necessity involved: that "goodness" may somehow be compelled to create a universe because it is good that it does so.

If one is prepared to go along with the idea that the universe does not exist reasonlessly, and if for convenience we label the reason God (whether one has in mind a person, a creative force, an ethical requirement, or some concept not yet formulated), then the first question to tackle is: in what sense might God be said to be responsible for the laws of physics (and other contingent features of the world)? For this notion to have any meaning at all, God must somehow *select* our world from the many alternatives. There has to be some element of choice involved. Some possible universes have to be discarded. So what sort of God would this be? By assumption, he would be rational. There is no point in invoking an irrational God; we might as well accept an irrational universe as it is. He should also be omnipotent. If God were not omnipotent, then his power would be limited in some way. But what would constrain this power? We should want to know in turn how this limitation originated, and what determined the form of the constraints: exactly what God was and was not allowed to do. (Notice that even an omnipotent God is subject to the constraints of logic. God couldn't make a square a circle, for example.) By similar reasoning, God would have to be perfect, for what agency would produce any defects? He would also have to be omniscient—that is, he would need to be aware of all the logically possible alternatives—so that he would be in a position to make a rational choice.

The Best of All Possible Worlds?

Leibniz developed the above argument in detail as an attempt to prove, on the basis of the rationality of the cosmos, that such a God exists. He concluded from this cosmological argument that a rational, omnipotent, perfect, omniscient being must inevitably choose the best of all possible worlds. The reason? If a perfect God knowingly selected a world that was less than perfect, that would be irrational. We would demand an explanation for the peculiar choice. But what possible explanation could there be?

The notion that ours is the best of all possible worlds has not commended itself to many people. Leibniz (in the guise of Dr. Pangloss) was savagely lampooned by Voltaire on this point: "O Dr. Pangloss! If this is the best of all possible worlds, what must the others be like?" The objection usually centers on the problem of evil. We can imagine a world in which, for example, there is no pain and suffering. Would that not be a better world?

Leaving ethical issues aside, there could still be some physical sense in which ours is the best of all possible worlds. One is certainly struck by the immense richness and complexity of the physical world. Sometimes it seems as if nature were "going out of its way" to produce an interesting and fruitful universe. Freeman Dyson has attempted to capture this property in his principle of maximum diversity: the laws of nature and the initial conditions are such as to make the universe as interesting as possible. Here "best" is interpreted as "richest," in the sense of greatest variety and complexity of physical systems. The trick is to make this mathematically precise in some way.

Recently mathematical physicists Lee Smolin and Julian Barbour have advanced an imaginative proposal for how this might be achieved. They conjecture that there is a fundamental principle of nature that makes the universe maximally varied. This means that things have arranged themselves so as to produce the greatest variety, in some sense to be defined precisely. Leibniz proposed that the world exhibits the maximum variety subject to the greatest degree of order. Impressive though this sounds, it doesn't add up to much unless it can be given a clear mathematical meaning. Smolin and Barbour make a start on this, albeit in a modest manner. They define "variety" for the simplest conceivable system: a collection of dots joined by a network of lines, like a map of airline routes. Mathematicians call this a "graph." The dots and lines don't have to correspond to real objects in real space, they just represent some sort of abstract interconnectedness that can be studied in its own right. Obviously there will be simple graphs and complicated graphs, depending on the way the connecting lines are put in. It is possible to find graphs that are in some well-defined sense the most varied in arrangement when looked at from all the different sites (dots). The trick is to relate all this to the real world. What are these dots and lines? The suggestion is they are some sort of abstract representation of particles in three-dimensional space, and that notions like the distance between particles might emerge naturally from the graph relationships. At this stage the idea remains pretty sketchy, but it shows at least the sort of things theorists could do to widen their horizons in their approach to the nature of physical laws.

Other forms of optimization can be imagined, different ways in which ours might be the best of all possible worlds. I mentioned that the laws of physics are like a cosmic code, a "message" cryptically buried in the data of our observations. John Barrow has speculated that the particular laws of our universe may represent some sort of optimal coding. Now, most of what scientists know about codes and information transmission stems from the pioneering wartime work of Claude Shannon, whose book on information theory became a classic. One of the problems Shannon addressed was the effect on a message of a noisy communication channel. We all know how noise on a telephone line can make conversation difficult; quite generally, noise degrades information. But you can get around the problem by coding the message with suitable redundancy. This is the principle behind some modern telecommunications systems. Barrow extends the idea to the laws of nature. Science is, after all, a dialogue with nature. When we conduct experiments, we are in a sense interrogating nature. Moreover, the information we get back is never pristine; it is degraded by all sorts of "noise" called experimental error, arising from many factors. But as I have stressed, nature's information is not in plaintext; it is in code. Barrow's proposal is that this "cosmic code" might be specially structured for optimal information transfer in analogy to Shannon's theory: "In order to realize this promise of arbitrarily high signalling fidelity the message must be encoded in a special way. . . . In some strange metaphorical way Nature seems to be 'encrypted' in one of those expedient forms."[12] This might explain our remarkable success at decoding the message and uncovering sweeping laws.

Another type of optimization related to the mathematical form of the laws of nature concerns their oft-cited simplicity. Einstein summed it up when he wrote: "Our experience hitherto justifies us in believing that nature is the realization of the simplest conceivable mathematical ideas."[13] This is certainly puzzling. "It is enigma enough that the world is described by mathematics," writes Barrow, "but by *simple* mathematics, of the sort that a few years energetic study now produces familiarity with, this is a mystery within an enigma."[14] So do we live in the best of all possible worlds in the sense that it has the simplest mathematical description? Earlier in this chapter I gave the reasons why I think not. What about the simplest possible world that permits the existence of biological complexity? Again, as I have already explained, I think the answer is no, but this is at least a conjecture which is open to scientific investigation. We can write down the equations of physics and then tinker with them a bit to see what difference it makes. In this way theorists can construct artificial-model universes to test mathematically whether they can support life. Considerable effort has gone into studying this question. Most investigators conclude that the existence of complex systems, especially biological systems, is remarkably sensitive to the form of the laws of physics, and that in some cases the most minute changes to the laws are sufficient to wreck the chances of life arising, at least in the form we know it. This topic goes under the name of the Anthropic Principle, because it relates our existence as observers of the universe to the laws and conditions of the universe. I shall return to it in chapter 8.

Of course, demanding that the laws admit conscious organisms may in any case be excessively chauvinistic. There could be many ways in which the laws are special, such as possessing all sorts of mathematical properties of which we may still be unaware. There are many obscure quantities that might be maximized or minimized by these particular laws. We just don't know.

Beauty as a Guide to Truth

So far I have dwelt on mathematics. But perhaps the laws distinguish themselves in other, more subtle ways, such as by their aesthetic value. It is widely believed among scientists that beauty is a reliable guide to truth, and many advances in theoretical physics have been made by the theorist demanding mathematical elegance of a new theory. Sometimes, where laboratory tests are difficult, these aesthetic criteria are considered even more important than experiment. Einstein, when discussing an experimental test of his general theory of relativity, was once asked what he would do if the experiment didn't agree with the theory. He was unperturbed at the prospect. "So much the worse for the experiment," he retorted. "The theory is right!" Paul Dirac, the theoretical physicist whose aesthetic deliberations led him to construct a mathematically more elegant equation for the electron, which then led to the successful prediction of the existence of antimatter, echoed these sentiments when he judged that "it is more important to have beauty in one's equations than to have them fit experiment."

Mathematical elegance is not an easy concept to convey to those unfamiliar with mathematics, but it is keenly appreciated by professional scientists. Like all aesthetic value-judgments, however, it is highly subjective. Nobody has yet invented a "beauty meter" that can measure the aesthetic value of things without referring to human criteria. Can one really say that certain mathematical forms are intrinsically more beautiful than others? Perhaps not. In which case it is very odd that beauty is such a good guide in science. Why should the laws of the universe seem beautiful to humans? No doubt there are all sorts of biological and psychological factors at work in framing our impressions of what is beautiful. It is no surprise that the female form is attractive to men, for example, and the curvaceous lines of many beautiful sculptures, paintings, and architectural structures doubtless have sexual referents. The structure and operation of the brain may also dictate what is pleasing to the eye or ear. Music may reflect cerebral rhythms in some fashion. Either way, though, there is something curious here. If beauty is entirely biologically programmed, selected for its survival value alone, it is all the more surprising to see it re-emerge in the esoteric world of fundamental physics, which has no direct connection with biology. On the other hand, if beauty is more than mere biology at work, if our aesthetic appreciation stems from contact with something firmer and more pervasive, then it is surely a fact of major significance that the fundamental laws of the universe seem to reflect this "something."

In chapter 6 I discussed how many distinguished scientists have expressed the feeling that their inspiration has come from some sort of mental contact with a Platonic realm of mathematical and aesthetic forms. Roger Penrose in particular is frank about his belief in the creative mind "breaking through" into the Platonic realm to glimpse mathematical forms which are in some way beautiful. Indeed, he cites beauty as a guiding principle in much of his mathematical work. This may appear surprising to readers who have an image of mathematics as an impersonal, cold, dry, and rigorous discipline. But as Penrose explains: "Rigorous argument is usually the *last* step! Before that, one has to make many guesses, and for these, aesthetic convictions are enormously important."[15]

Is God Necessary?

> "Man has two eyes
> One only sees what moves in fleeting time
> The other
> What is eternal and divine"
>
> *The Book of Angelus Silesius*

Moving on from the question of whether, and in what sense, we might be living in the best of all possible worlds, we have to confront a still deeper problem. Put simply, if the universe really has an explanation and it can't explain itself, then it must be explained by something outside itself—e.g., God. But what, then, explains God? This age-old "who made God" conundrum is in danger of pitching us into an infinite regress. The only escape, it would seem, is to assume that God can somehow "explain himself," which is to say that God is a *necessary* being in the technical sense that I explained at the beginning of this chapter. More precisely, if God is to supply the sufficient reason for the universe, then it follows that he himself must be a necessary being, for, if God were contingent, then the chain of explanation would still not have terminated, and we would want to know what were the factors beyond God on which his existence and nature depended. But can we make sense of the notion of a necessary being, a being that contains entirely within itself the reason for its own existence? Many philosophers have argued that the idea is incoherent or meaningless. Certainly human beings are not able to comprehend the nature of such a being. But that in itself does not mean that the notion of a necessary being is self-contradictory.

In coming to grips with the concept of a necessary being, one can start by asking whether there is anything that is necessarily the case. As an appetite whetter, consider the statement: "There is at least one true proposition." Call this proposition A. Is A necessarily true? Suppose I contend that A is false. Call this proposition B: "A is false."

177

But if A is false, so is B, because B is a proposition, and if A is false there are no true propositions. So A must be true. It is therefore logically impossible for there to exist no true propositions.

If there exist necessary propositions, then the notion of a necessary being is not obviously absurd. The traditional God of Christian theology, developed in large part by Saint Thomas Aquinas in the thirteenth century, is a necessary, timeless, immutable, perfect, unchanging being on which the universe depends utterly for its existence, but who in contrast is completely unaffected by the existence of the universe. Although the demands of rationality seem to compel us toward such an image of God as the ultimate explanation of the world, there is a serious difficulty about relating this God to a contingent, changing universe, especially a universe containing beings with free will. As the atheist philosopher A. J. Ayer once expressed it, from necessary propositions only necessary propositions follow.

This devastating contradiction has lurked at the heart of Western theology ever since Plato. For Plato, as we have seen, the very concept of "rational" was tied to the existence of an abstract world of eternal, unchanging, perfect Forms which for him represented the only true reality. And located in this immutable realm was the ultimate object of knowledge, the Good. By contrast, the directly perceived world of material things is forever in a state of flux. The relationship between the eternal world of Forms and the changing world of matter is then deeply problematical. As I explained in chapter 1, Plato proposed the existence of a Demiurge who is located within time, and who fashions matter as best he can using the Forms as a blueprint. But this naïve attempt to reconcile the changing and the Changeless, the imperfect and the Perfect, only serves to underline the seriousness of the conceptual paradox which dogs all explanations of contingency.

It is important to understand that the paradox is more than just a technicality of theological debate; it is an inevitable consequence of certain rational methods of explanation. Descartes and his followers have sought to root our experience of the world in the bedrock of intellectual certainty. If we adhere to that tradition, then in our search for the most secure form of knowledge we are inevitably led to timeless concepts such as mathematics and logic, because real truth, by definition, cannot change with time. And the dependability of this abstract realm is assured because its elements are anchored to each other by the certainty of logical necessity. Yet the very world of experiences we seek to explain is time-dependent and contingent.

The tension which this mismatch generates pervades science as surely as it pervades religion. We see it in the endless confusion that surrounds attempts to reconcile eternal laws of physics with the existence of an "arrow of time" in the universe. We see it in fierce debates about how to square progressive biological evolution with directionless mutation. And we see it in the clash of paradigms that accompanies the recent work on self-organizing systems, the hostile reception to which indicates deep-seated cultural prejudices.

The unique contribution of Christian thinking to this tension is the doctrine of creation *ex nihilo,* which I introduced in chapter 2. Here was a brave attempt to break out of the paradox by proposing a timeless, necessary being who brings into existence (not within time) a material universe by divine power as an act of free choice. By declaring that the creation is something other than the Creator, something which God did not have to create but chose to, Christians escaped from the strictures of the alternative scheme of divine emanation, wherein the physical universe issues directly from God's essence and is thus imprinted with his necessary properties. The key element that is introduced here is that of the divine Will. By definition, free will entails contingency, because we say that a choice is free only if it could have been otherwise. So, if God is endowed with a freedom to choose between alternative possible worlds, the contingency of the actual world is explained. Yet the demand for intelligibility is preserved by attributing to God a rational nature, thereby ensuring a rational choice.

This seems to be real progress. It appears as if creation *ex nihilo* resolves the paradox of how a changing, contingent world can be explained by a timeless necessary being. Unfortunately, in spite of the attention of generations of philosophers and theologians to develop this idea into a coherent scheme, major obstacles remain. The chief one is to understand why God chose to create this particular world rather than some other. When human beings choose freely, their choice is colored by their nature. So what can be said about God's nature? Presumably that is fixed by his necessity. We don't want to contend with the possibility that there could be many different types of God, for then we would have gained nothing by invoking God in the first place. We would be left with the problem of explaining why *that* particular God existed rather than some other. The whole idea of invoking God as a *necessary* being is to ensure that he is unique: his nature could not have been otherwise. But if God's nature is fixed by his necessity, could he have chosen to create a different universe? Only if his choice was not rational at all, but whimsical, the theistic equivalent of tossing a coin. But in that case existence is arbitrary, and we might as well be content with an arbitrary universe and leave it at that.

The philosopher Keith Ward has made a detailed study of the clash between God's necessity and the contingency of the world. He summarizes the essential dilemma as follows:

> First of all, if God really is self-sufficient, as the axiom of intelligibility seems to require him to be, how can it come about that he creates a world at all? It seems an arbitrary and pointless exercise. On the other hand, if God really is a necessary and immutable being, how can he have a free choice; surely all that he does will have to be done of necessity and without any possibility of alteration? The old dilemma—either God's acts are necessary and therefore not free (could not be otherwise), or they are free and therefore arbitrary (nothing determines what they shall be)—has been sufficient to impale the vast majority of Christian philosophers down the ages. [16]

The problem is, whichever way you cut the cake, you come back to the same basic difficulty, that the truly contingent cannot arise from the wholly necessary:

> If God is the creator or cause of a contingent world, he must be contingent and temporal; but if God is a necessary being, then whatever he causes must be necessarily and changelessly caused. On this rock both interpretations of theism founder. The demands of intelligibility require the existence of a necessary, immutable, eternal being. Creation seems to demand a contingent, temporal God, who interacts with creation and is, therefore, not self-sufficient. But how can one have both? [17]

And elsewhere:

> How can a being which is necessary and immutable have the power to do everything? Being necessary, it cannot do anything other than it does. Being immutable, it cannot do anything new or original. . . . Even if creation can be seen as a timeless Divine act, the real difficulty remains, that, since the being of God is wholly necessary, it will be a necessary act, which could not have been otherwise in any respect.

This view is still in tension with a central strand of the Christian tradition: namely, that God need not have created any universe, and that he need not have created precisely this universe. How can a necessary being be free in any way?[18]

The same point is made by Schubert Ogden:

Theologians usually tell us that God creates the world freely, as the contingent or non-necessary world of our experience discloses it to be. . . . At the same time, because of their fixed commitment to the assumptions of classical metaphysics, theologians usually tell us that God's act of creation is one with his eternal essence, which is in every respect necessary, exclusive of all contingency. Hence, if we take them at their word, giving full weight to both of their assertions, we at once find ourselves in the hopeless contradiction of a wholly necessary creation of a wholly contingent world.[19]

Volumes have been written by theologians and philosophers in an attempt to break out of this glaring and persistent contradiction. For reasons of space I shall limit myself to discussing one particular, and rather obvious, escape route.

A Dipolar God and Wheeler's Cloud

As we have seen, Plato confronted the paradox of necessity versus contingency by proposing two gods, one necessary, the other contingent: the Good and the Demiurge. Perhaps the demands of monotheism can be met by arguing that this situation could be legitimately described as really two complementary aspects of a single "dipolar" God. This is the position adopted by proponents of what is known as "process theology."

Process thought is an attempt to view the world not as a collection of objects, or even as a set of events, but as a *process* with a definite directionality. The flux of time thus plays a key role in process philosophy, which asserts the primacy of *becoming* over *being*. In contrast to the rigid mechanistic view of the universe that arose from the work of Newton and his associates, process philosophy stresses the openness and indeterminism of nature. The future is not implicit in the present:

there is a choice of alternatives. Thus nature is attributed a sort of freedom which was absent in the clockwork universe of Laplace. This freedom comes about through the abandonment of reductionism: the world is more than the sum of its parts. We must reject the idea that a physical system, such as a rock or a cloud or a person, is *nothing but* a collection of atoms, and recognize instead the existence of many different levels of structure. A human being, for example, certainly is a collection of atoms, but there are many higher levels of organization that are missed by this meager description and which are essential for defining what we mean by the word "person." By viewing complex systems as a hierarchy of organizational levels, the simple "bottom-up" view of causality in terms of elementary particles interacting with other particles must be replaced by a more subtle formulation in which higher levels can act downward upon lower levels too. This serves to introduce elements of teleology, or purposive behavior, in the affairs of the world. Process thinking leads naturally to an organismic or ecological view of the universe, reminiscent of Aristotle's cosmology. Ian Barbour describes the process vision of reality as the view that the world is a community of interdependent beings rather than a collection of cogs in a machine.

Although strands of process thought have a long-established place in the history of philosophy, it is only in recent years that process thinking has become fashionable in science. The rise of quantum physics in the 1930s put paid to the idea of the universe as a deterministic machine, but the more recent work on chaos, self-organization, and nonlinear-systems theory has been more influential. These topics have forced scientists to think more and more about *open* systems, which are not rigidly determined by their component parts because they can be influenced by their environment. Typically, complex, open systems can have incredible sensitivity to external influences, and this makes their behavior unpredictable, bestowing upon them a type of freedom. What has come as a surprise is that *open systems can also display ordered and lawlike behavior in spite of being indeterministic and at the mercy of seemingly random outside perturbations.* There appear to exist general organizing principles that supervise the behavior of complex systems at higher organizational levels, principles that exist alongside the laws of physics (which operate at the bottom level of individual particles). These organizing principles are consistent with, but cannot be reduced to, or derived from, the laws of physics. Scientists have thus rediscovered the crucial quality of *contingent order*. A more detailed discussion of these topics can be found in *The Cosmic Blueprint* and *The Matter Myth*.

Process thought was introduced into theology by the mathematician and philosopher Alfred North Whitehead, who was co-author with Bertrand Russell of the seminal work *Principia Mathematica*. Whitehead proposed that physical reality is a network linking what he termed "actual occasions," these being more than mere events, for they are invested with a freedom and an internal experience that are lacking in the mechanistic view of the world. Central to Whitehead's philosophy is that God is responsible for ordering the world, not through direct action, but by providing the various potentialities which the physical universe is then free to actualize. In this way, God does not compromise the essential openness and indeterminism of the universe, but is nevertheless in a position to encourage a trend toward good. Traces of this subtle and indirect influence may be discerned in the progressive nature of biological evolution, for example, and the tendency for the universe to self-organize into a richer variety of ever more complex forms. Whitehead thus replaces the monarchical image of God as omnipotent creator and ruler to that of a participator in the creative process. He is no longer self-sufficient and unchanging, but influences, and is influenced by, the unfolding reality of the physical universe. On the other hand, God is not thereby completely embedded in the stream of time. His basic character and purposes remain unchanging and eternal. In this way, timelessness and temporality are folded into a single entity.

Some people claim that a "dipolar" God can also combine necessity and contingency. Achieving this, however, means giving up any hope that God might be *simple* in his divine perfection, as Aquinas supposed. Keith Ward, for example, has proposed a complex model for God's nature, some parts of which might be necessary, others contingent. Such a God, though necessarily existent, is nevertheless changed by his creation, and by his own creative action, which includes an element of openness or freedom.

I confess I have had to struggle hard to understand the philosophical convolutions needed to justify a dipolar God. Help came, however, from an unexpected source: quantum physics. Let me reiterate yet again the central message of quantum uncertainty. A particle such as an electron cannot have a well-defined position and a well-defined momentum at the same time. You can make a measurement of position and obtain a sharp value, but in this case the value of the momentum is completely uncertain, and vice versa. For a general quantum state, it is impossible to say in advance what value will be obtained by a measurement: only probabilities can be assigned. Thus, when one is making a position measurement on such a state, a range of outcomes is available. The system is therefore indeterministic—one might say, free to choose among a range of possibilities—and the actual outcome is contingent. On the other hand, the experimenter determines whether the measurement shall be of position or momentum, so the class of alternatives (i.e., a range of position values or a range of momentum values) is decided by an external agent. As far as the electron is concerned, the nature of the alternatives is fixed necessarily, whereas the actual alternative adopted is contingent.

To make this clearer, let me retell a famous parable due to John Wheeler. One day Wheeler was unwittingly subjected to a variant of the game of twenty questions. Recall that, in the conventional game, the players agree on a word and the subject tries to guess the word by asking up to twenty questions. Only yes-no answers can be given. In the variant version, Wheeler began by asking the usual questions: Is it big? Is it living? etc. At first the answers came quickly, but as the game went on they became slower and more hesitant. Eventually he tried his luck: "Is it a cloud?" The answer came back: "Yes!" Then everyone burst out laughing. The players revealed that to trick Wheeler no word had been chosen in advance. Instead they agreed to answer his questions purely at random, subject only to consistency with previous answers. Nevertheless, an answer was obtained. This obviously contingent answer was not determined in advance, but neither was it arbitrary: its nature was decided in part by the questions Wheeler chose to ask, and in part by pure chance. In the same way, the reality exposed by a quantum measurement is decided in part by the questions the experimenter puts to nature (i.e., whether to ask for a definite position or a definite momentum) and in part by chance (i.e., the uncertain nature of the values obtained for these quantities).

Let us now return to the theological analogue. This mixture of contingency and necessity corresponds to a God who necessarily determines what alternative worlds are available to nature, but who leaves open the freedom of nature to choose from among the alternatives. In process theology the assumption is made that the alternatives are necessarily fixed in order to achieve a valued end result—i.e., they direct or encourage the (otherwise unconstrained) universe to evolve toward something good. Yet within this directed framework there remains openness. The world is therefore neither wholly determined nor arbitrary but, like Wheeler's cloud, an intimate amalgam of chance and choice.

Does God Have to Exist?

So far in this chapter I have been tracing the consequences of the cosmological argument for the existence of God. This argument does not attempt to establish that God's existence is a *logical* necessity. One can certainly imagine that neither God nor the universe existed, or that the universe existed without God. On the face of it there does not seem to be any logical contradiction in either state of affairs. So, even if a case can be made that the concept of a necessary being makes sense, it does not follow that such a being exists, still less has to exist.

The history of theology is not, however, without attempts to prove that God's nonexistence is logically impossible. This argument, known as the "ontological argument," goes back to Saint Anselm, and runs something like this. God is defined to be the greatest conceivable thing. Now, a really existing thing is obviously greater than the mere idea of that thing. (A real person—for example, the famous Fabian of Scotland Yard—is greater than a fictional character, such as Sherlock Holmes.) Therefore, a really existing god is greater than an imaginary god. But as God is the greatest conceivable thing, it follows that he must exist.

The fact that the ontological argument reeks of logical trickery belies its philosophical force. It has in fact been taken very seriously by many philosophers over the years, including briefly by the atheistic Bertrand Russell. Nevertheless, even theologians have not generally been prepared to defend it. One problem lies with the treatment of "existence" as if it were a property of things, like mass or color. Thus the argument obliges one to compare the concepts of gods-that-really-exist and gods-that-don't-really-exist. But existence is not the sort of attribute to be placed alongside normal physical properties. I can meaningfully talk about having five little coins and six big coins in my pocket, but what does it mean for me to say that I have five existing coins and six nonexistent coins?

A further problem with the ontological argument is the requirement that God explain the world. It is not sufficient for there to exist a logically necessary being who is in no way related to the world. But it is hard to see how a being who exists in the realm of pure logic can explain the contingent properties of the world. The ontological argument relies on what philosophers call "analytic propositions." An analytic proposition is one whose truth (or otherwise) follows purely from the meanings of the words involved. Thus "All bachelors are men" is an analytic proposition. Propositions that do not fall into this class are called "synthetic," because they make connections between things that are not related merely by definition. Now, physical theories always involve synthetic propositions, because they make statements about the facts of nature that can be tested. The success of mathematics in describing nature, especially the underlying laws, can give the impression (defended by some, as we have seen) that there is nothing more to the world than mathematics, and that this mathematics is in turn nothing more than definitions and tautologies—i.e., analytical propositions. I believe this line of thinking to be badly misconceived. However hard you try, you cannot derive a synthetic proposition from an analytic proposition.

Immanuel Kant was an opponent of the ontological argument. He maintained that, if there were to be any meaningful metaphysical statements, then there must exist propositions that are necessarily true other than by virtue of mere definition. In chapter 1 I explained that Kant believed we possess some knowledge *a priori*. Thus Kant asserted that there have to be some true synthetic *a priori* propositions for any process of thought concerned with an objective world. These synthetic *a prioris* would have to be true independently of the contingent features of the world—i.e., they must be true in any world. Unfortunately, philosophers have yet to be convinced that there are any necessary synthetic *a priori* propositions.

Even if there are no synthetic propositions that are necessary, there may be some that are unobjectionable. One could imagine that a set of such propositions might explain the contingent features of the world, such as the form of the laws of physics. Many people might be satisfied with this. Physicist David Deutsch argues that, "instead of trying to get 'something for nothing,' a synthetic proposition from an analytic," we should introduce into physics at the fundamental level synthetic propositions "which have to be postulated anyway for some reason outside physics." He goes on to suggest an example.

One thing that we always tacitly assume a priori in the search for any physical theory is that the physical process of that theory becoming known and expressed is not in itself forbidden by the theory. No physical principle that we can know can itself forbid our knowing it. That every physical principle must satisfy this highly restrictive property is a synthetic a priori proposition, not because it is necessarily true, but because we cannot help assuming it to be true in seeking to know the principle. [20]

John Barrow also suggests that there are certain necessary truths about any world that can be observed. He cites the various Anthropic Principle arguments which seek to demonstrate that conscious biological organisms can arise only in a universe in which the laws of physics have a certain special form: "These 'anthropic' conditions . . . point us towards certain properties that the Universe must possess a priori, but which are non-trivial enough to be counted as synthetic. The synthetic a priori begins to look like the requirement that every knowable physical principle that forms part of the 'Secret of the Universe' must not forbid the possibility of our knowing it."[21]

Keith Ward argues that we could define a broader notion of logical necessity. For example, consider the statement: "Nothing can be red and green all over." Is this statement necessarily true? Suppose I assert it is false. My assertion is not obviously self-contradictory. Nevertheless, it may still be false in all possible worlds: that is not the same as saying it is logically self-contradictory in a formal sense. The assumption that the statement is true is, in Deutsch's words, "something that we would make anyway." Perhaps, then, the statement "God does not exist" falls into this category. The statement might not contradict the axioms of some formal scheme of propositional logic, yet it might be the case that the statement is false in all possible worlds.

Finally, mention should be made of Frank Tipler's application of the ontological argument to the universe itself (as opposed to God). Tipler attempts to circumvent the objection that "existence" is not a property of something by defining existence in an unusual way. In chapter 5 we saw how Tipler defends the notion that computer-simulated worlds are every bit as real for the simulated beings as our world is to us. But he points out that a computer program is, in essence, nothing but a mapping of one set of symbols or numbers into another set. One might consider that all possible mappings—hence all possible computer programs—exist in some abstract Platonic sense. Among these programs will be many (probably an infinity) that represent simulated universes. The question is, which among the many possible computer simulations correspond to "physically existing" universes? To use Hawking's terms, which ones have fire breathed into them? Tipler proposes that those simulations "which are sufficiently complex to contain observers—thinking, feeling beings—as subsimulations" are the ones that exist physically, at least as far as the simulated beings are concerned. Furthermore, these simulations *necessarily* exist as a consequence of the logical requirements of the mathematical operations involved in the mappings. Therefore, concludes Tipler, our universe (and a great many others) *must* exist as a consequence of logical necessity.

The Options

So what are we to conclude? If the reader is bewildered after this little tour of philosophy, so is the author. It seems to me that the ontological argument is an attempt to define God into existence from nothing, and, as such, in a strictly logical sense it cannot succeed. You can't get out of a purely deductive argument more than you put into the premises. At best the argument can demonstrate that, if a necessary being is possible, then he must exist. God could only fail to exist if the concept of a necessary being is incoherent. I can accept that. But the argument fails to demonstrate the strictly formal impossibility of God's nonexistence. On the other hand, if the ontological argument is augmented with an extra assumption or assumptions, then it could be successful. Now, what if these extra assumptions (which would necessarily be synthetic) were limited to presuppositions necessary for the existence of rational thought? We could then conclude that the activity of rational inquiry would indeed be able to establish God's existence by reason alone. This suggestion is mere speculation, but Keith Ward for one is prepared to keep an open mind about it: "It is not absurd to think that by analysis of the notions of 'perfection,' 'being,' 'necessity' and 'existence,' one might find that a presupposition of their objective applicability to the world is the existence of an object of a certain type."[22]

What of the cosmological argument? If we accept the world's contingency, then one possible explanation is the existence of a transcendent God. We then have to confront the issue of whether God is necessary or contingent. If God is simply contingent, have we really gained anything by invoking him, for his own existence and qualities remain unexplained? It is possible that we have. It may be that the hypothesis of a God provides a simplifying and unifying description of reality that improves on the "package" acceptance of a list of laws and initial conditions. The laws of physics may be able to take us only so far, and we could then seek a deeper level of explanation. The philosopher Richard Swinburne, for example, has argued that it is simpler to posit the existence of an infinite mind than to accept, as a brute fact, the existence of this contingent universe. In this case belief in God is largely a matter of taste, to be judged by its explanatory value rather than logical compulsion. Personally I feel more comfortable with a deeper level of explanation than the laws of physics. Whether the use of the term "God" for that deeper level is appropriate is, of course, a matter of debate.

Alternatively, one could embrace the classical theistic position and argue that God is a necessary being who creates a contingent universe as an act of his free will. That is, God has no choice about his own existence and qualities, but he does have a choice about the universe he creates. As we have seen, this position is fraught with philosophical difficulties, though it may be that some resolution can be found. Most attempted resolutions descend into a quagmire of linguistic niceties concerning the many definitions of "necessity," "truth," and so on, and many seem to peter out with the frank acceptance of mystery. But the dipolar concept of God, in which distinction is made between God's necessary nature and his contingent actions in the world, though having the disadvantage of complexity, comes closest to circumventing these problems.

What seems to come through such analyses loud and clear is the fundamental incompatibility of a completely timeless, unchanging, necessary God with the notion of creativity in nature, with a universe that can change and evolve and bring forth the genuinely new, a universe in which there is free will. You can't really have it both ways. Either God fixes everything, including our own behavior, in which case free will is an illusion—"The plan of predestination is certain," wrote Aquinas—or things happen over which God either has no control, or has voluntarily relinquished control.

Before we leave the problem of contingency, something should be said about the so-called many-universes theory. According to this idea, currently popular with some physicists, there is not just one physical universe, but an infinity of them. All these universes somehow coexist "in parallel," each differing from the others, perhaps only slightly. It is conceivable that things could be arranged such that every sort of universe that is possible exists in this infinite set. If you want a universe, say, with an inverse-cube rather than inverse-square law of gravity, well, you will find one there somewhere. Most of these universes will not be inhabited, because the physical conditions therein will not be suitable for the formation of living organisms. Only those universes in which life can form and flourish to the point that conscious individuals arise will be observed. The rest go unseen. Any given observer will observe only a particular universe, and will not be directly aware of the others. That particular universe will be strongly contingent. Nevertheless, the question "Why this universe?" is no longer relevant, because all possible universes exist. The set of all universes taken together is not contingent.

Not everybody is happy with the many-universes theory. To postulate an infinity of unseen and unseeable universes just to explain the one we do see seems like a case of excess baggage carried to the extreme. It is simpler to postulate one unseen God. This is the conclusion also reached by Swinburne:

> The postulation of God is the postulation of *one* entity of a simple kind. . . . The postulation of the actual existence of an infinite number of worlds, between them exhausting all the logical possibilities . . . is to postulate complexity and non-prearranged coincidence of infinite dimensions beyond rational belief.[23]

Scientifically the many-universes theory is unsatisfactory because it could never be falsified: what discoveries would lead a many-worlder to change her/his mind? What can you say to convince somebody who denies the existence of these other worlds? Worse still, you can use many worlds to explain anything at all. Science becomes redundant. The regularities of nature would need no further investigation, because they could simply be explained as a selection effect, needed to keep us alive and observing. Furthermore, there is something philosophically unsatisfactory about all those universes that go unobserved. To paraphrase Penrose, what does it mean to say that something exists that can never in principle be observed? I shall have more to say on this topic in the next chapter.

A God Who Plays Dice

I concede that one cannot prove the world to be rational. It is certainly possible that, at its deepest level, it is absurd, and we have to accept the existence and features of the world as brute facts that could have been otherwise. Yet the success of science is at the very least strong circumstantial evidence in favor of the rationality of nature. In science, if a certain line of reasoning is successful, we pursue it until we find it to fail.

In my own mind I have no doubts at all that the arguments for a necessary world are far shakier than the arguments for a necessary being, so my personal inclination is to opt for the latter. Yet I also believe that there remain severe difficulties relating this timeless, necessary being to the changing, contingent world of experience, for the reasons I have discussed. I don't believe that these difficulties can be separated from the various unresolved puzzles that exist anyway concerning the nature of time, the freedom of the will, and the notion of personal identity. Nor is it obvious to me that this postulated being who underpins the rationality of the world bears much relation to the personal God of religion, still less to the God of the bible or the Koran.

Though I have no doubts at all as to the rationality of nature, I am also committed to the notion of a creative cosmos, for reasons I have set out in my book *The Cosmic Blueprint*. And here we inevitably encounter the paradox of reconciling being and becoming, the changing and the eternal. This can only be done by a compromise. The compromise is called "stochasticity." A stochastic system is, roughly speaking, one which is subject to unpredictable and random fluctuations. In modern physics, stochasticity enters in a fundamental way in quantum mechanics. It is also inevitably present when we deal with open systems subject to chaotic external perturbations.

In modern physical theory, rationality is reflected in the existence of fixed mathematical laws, and creativity is reflected in the fact that these laws are fundamentally statistical in form. To use once more Einstein's well-worn phrase, God plays dice with the universe. The intrinsically statistical character of atomic events and the instability of many physical systems to minute fluctuations, ensures that the future remains open and undetermined by the present. This makes possible the emergence of new forms and systems, so that the universe is endowed with a sort of freedom to explore genuine novelty. I thus find myself closely in tune with process thought, as described earlier in this chapter.

I am aware that introducing stochasticity at a fundamental level into nature implies a partial abandonment of the principle of sufficient reason. If there is a genuine stochasticity in nature, then the outcome of any particular "throw of the die" is genuinely undetermined by anything, which is to say that there is no reason why, in that particular case, that particular result was forthcoming. Let me give an example. Imagine an electron colliding with an atom. Quantum mechanics tells us that there is, say, equal probability that the electron will deflect to the left as to the right. If the statistical nature of quantum events is truly inherent, and not merely a result of our ignorance, then, if the electron actually deflects to the left, there will be no reason whatever why it has done so, rather than deflect to the right.

Is this not to admit an element of irrationality into the world? Einstein thought so ("God does not play dice with the universe!"). This was why he could never accept that quantum mechanics gives a complete description of reality. But one man's irrationality is another person's creativity. And there is a difference between stochasticity and anarchy. The development of new forms and systems is subject to general principles of organization that guide and encourage, rather than compel, matter and energy to develop along certain predetermined pathways of evolution. In *The Cosmic Blueprint* I used the word "predestination" to refer to these general tendencies, to distinguish it from "determinism" (which is the sense in which Aquinas uses the term). For those, such as process theologians, who choose to see God's guiding hand rather than genuine spontaneity in the way the universe develops creatively, then stochasticity can be regarded as an efficient device through which divine intentions can be carried out. And there is no need for such a God to interfere directly with the course of evolution by "loading the dice," a suggestion I mentioned in passing in chapter 5. Guidance can be through the (timeless) laws of organization and information flow.

It might be objected that, if one is prepared to abandon the principle of sufficient reason at some stage, it can be abandoned elsewhere too.

If a particular electron "just happens" to deflect to the left, might it not be the case that the inverse-square law of gravitation, or the cosmic initial conditions, "just happen" to be the case? I believe the answer is no. The stochasticity inherent in quantum physics is fundamentally different in this respect. The condition of total disorder or randomness—the "fairness" of the quantum dice—is itself a law of a rather restrictive nature. Although each individual quantum event may be genuinely unpredictable, a collection of such events conforms with the statistical predictions of quantum mechanics. One might say that there is order in disorder. The physicist John Wheeler has stressed how lawlike behavior can emerge from the apparent lawlessness of random fluctuations, because even chaos can possess statistical regularities. The essential point here is that quantum events form an ensemble that we can observe. By contrast, the laws of physics and the initial conditions do not. It is one thing to argue that each event in a selection of chaotic processes just happens to be what it is, quite another to argue the same for an ordered process such as a law of physics.

So far in this philosophical excursion I have been largely concerned with logical reasoning. Little reference has been made to empirical facts about the world. On their own, the ontological and cosmological arguments are only a signpost for the existence of a necessary being. This being remains shadowy and abstract. If such a being exists, can we tell anything about his/her/its nature from an examination of the physical universe? This question brings me to the subject of design in the universe.

8

Designer Universe

HUMAN BEINGS have always been awestruck by the subtlety, majesty, and intricate organization of the physical world. The march of the heavenly bodies across the sky, the rhythms of the seasons, the pattern of a snowflake, the myriads of living creatures so well adapted to their environment—all these things seem too well arranged to be a mindless accident. There is a natural tendency to attribute the elaborate order of the universe to the purposeful workings of a Deity.

The rise of science served to extend the range of nature's marvels, so that today we have discovered order from the deepest recesses of the atom to the most distant galaxies. But science has also provided its own reasons for this order. No longer do we need theological explanations for snowflakes, or even for living organisms. The laws of nature are such that matter and energy can organize *themselves* into the complex forms and systems which surround us. Though it would be rash to claim that scientists understand everything about this self-organization, there seems to be no fundamental reason why, given the laws of physics, all known physical systems cannot be satisfactorily explained as the product of ordinary physical processes.

Some people conclude from this that science has robbed the universe of all mystery and purpose, and that the elaborate arrangement of the physical world is either a mindless accident or an inevitable consequence of mechanistic laws. "The more the universe seems comprehensible, the more it also seems pointless," believes physicist Steven Weinberg.[1] The biologist Jacques Monod echoes this dismal sentiment:

"The ancient covenant is in pieces: man at last knows that he is alone in the unfeeling immensity of the universe, out of which he has emerged only by chance. Neither his destiny nor his duty have been written down."[2]

Not all scientists, however, draw the same conclusions from the facts. Though accepting that the organization of nature can be explained by the laws of physics, together with suitable cosmic initial conditions, some scientists recognize that many of the complex structures and systems in the universe depend for their existence on the *particular form* of these laws and initial conditions. Furthermore, in some cases the existence of complexity in nature seems to be very finely balanced, so that even small changes in the form of the laws would apparently prevent this complexity from arising. A careful study suggests that the laws of the universe are remarkably felicitous for the emergence of richness and variety. In the case of living organisms, their existence seems to depend on a number of fortuitous coincidences that some scientists and philosophers have hailed as nothing short of astonishing.

The Unity of the Universe

There are several different aspects to this "too-good-to-be-true" claim. The first of these concerns the general orderliness of the universe. There are endless ways in which the universe might have been totally chaotic. It might have had no laws at all, or merely an incoherent jumble of laws that caused matter to behave in disorderly or unstable ways. Alternatively, the universe could have been extremely simple to the point of featurelessness—for example, devoid of matter, or of motion. One could also imagine a universe in which conditions changed from moment to moment in a complicated or random way, or even in which everything abruptly ceased to exist. There seems to be no logical obstacle to the idea of such unruly universes. But the real universe is not like this. It is highly ordered. There exist well-defined laws of physics and definite cause-effect relationships. There is a dependability in the operation of these laws. The course of nature continues always uniformly the same, to use David Hume's phrase. This causal order doesn't follow from logical necessity; it is a synthetic property of the world, and one for which we can rightly demand some sort of explanation.

The physical world does not merely display arbitrary regularities; it is ordered in a very special manner. As explained in chapter 5, the universe is poised interestingly between the twin extremes of simple regimented orderliness (like that of a crystal) and random complexity (as in a chaotic gas). The world is undeniably complex, but its complexity is of an *organized* variety. The states of the universe have "depth," to use the technical term introduced in chapter 5. This depth was not built into the universe at its origin. It has emerged from primeval chaos in a sequence of self-organizing processes that have progressively enriched and complexified the evolving universe. It is easy to imagine a world that, though ordered, nevertheless does not possess the right sort of forces or conditions for the emergence of significant depth.

There is another sense in which the order of the physical world is special. This concerns the general coherence and unity of nature, and the very fact that we can talk meaningfully about "the universe" at all as an all-embracing concept. The world contains individual objects and systems, but they are structured such that, taken together, they form a unified and consistent whole. For example, the various forces of nature are not just a haphazard conjunction of disparate influences. They dovetail together in a mutually supportive way which bestows upon nature a stability and harmony that are hard to capture mathematically but obvious to anyone who studies the world in depth. I have already tried to convey what I mean by this dovetailing consistency with the use of the crossword analogy.

It is particularly striking how processes that occur on a microscopic scale—say, in nuclear physics—seem to be fine-tuned to produce interesting and varied effects on a much larger scale—for example, in astrophysics. Thus we find that the force of gravity combined with the thermodynamical and mechanical properties of hydrogen gas are such as to create large numbers of balls of gas. These balls are large enough to trigger nuclear reactions, but not so large as to collapse rapidly into black holes. In this way, stable stars are born. Many large stars die in spectacular fashion by exploding as so-called supernovae. Part of the explosive force derives from the action of one of nature's most elusive subatomic particles—the neutrino. Neutrinos are almost entirely devoid of physical properties: the average cosmic neutrino could penetrate many light-years of solid lead. Yet these ghostly entities can still, under the extreme conditions near the center of a dying massive star, pack enough punch to blast much of the stellar material into space. This detritus is richly laced with heavy elements of the sort from which planet Earth is made. We can thus attribute the existence of terrestrial-like planets, with their huge variety of material forms and systems, to the qualities of a subatomic particle that might never have been discovered, so feeble is its action. The life cycles of stars provide just one example of the ingenious and seemingly contrived way in which the large-scale and small-scale aspects of physics are closely intertwined to produce complex variety in nature.

In addition to this coherent interweaving of the various aspects of nature, there is the curious uniformity of nature. Laws of physics discovered in the laboratory apply equally well to the atoms of a distant galaxy. The electrons that make the image on your television screen have exactly the same mass, charge, and magnetic moment as those on the moon, or at the edge of the observable universe. Furthermore, these qualities persist with no detectable change from one moment to the next. The magnetic moment of the electron, for instance, can be measured with ten-figure accuracy; even to such fantastic precision, no variation in this property has been found. There is also good evidence that the basic properties of matter cannot have varied much, even over the age of the universe.

As well as the uniformity of the laws of physics, there is also uniformity in the spatial organization of the universe. On a large scale, matter and energy are distributed extremely evenly, and the universe appears to be expanding at the same rate everywhere and in all directions. This means that an alien being in another galaxy would see very much the same sort of large-scale arrangement of things that we do. We share with other galaxies a common cosmography and a common cosmic history. As described in chapter 2, cosmologists have attempted to explain this uniformity using the so-called inflationary-universe scenario, which involves a sudden jump in the size of the universe shortly after its birth. This would have the effect of smoothing out any initial irregularities. It is important to realize, however, that explaining the uniformity in terms of a physical mechanism does nothing to lessen its specialness, for we can still ask why the laws of nature are such as to permit that mechanism to work. The point at issue is not the way in which the very special form came about, but that the world is so structured that it *has* come about.

Finally, there is the much-discussed simplicity of the laws. By this I mean that the laws can be expressed in terms of simple mathematical functions (like the inverse-square law). Again, we can imagine worlds in which there are regularities but of a very complicated sort requiring a clumsy combination of different mathematical factors. The charge that we develop our mathematics precisely to make the world look simple I have dealt with in chapter 6. I believe that the "unreasonable effectiveness" of mathematics in describing the world is an indication that the regularities of nature are of a very special sort.

Life Is So Difficult

I have tried to make a case that the existence of an orderly, coherent universe containing stable, organized, complex structures requires laws and conditions of a very special kind. All the evidence suggests that this is not just any old universe, but one which is remarkably well adjusted to the existence of certain interesting and significant entities (e.g., stable stars). In chapter 7 I explained how this feeling had been formalized by Freeman Dyson and others into something like a principle of maximum diversity.

The situation becomes even more intriguing when we take into account the existence of living organisms. The fact that biological systems have very special requirements, and that these requirements are, happily, met by nature, has been commented upon at least since the seventeenth century. It is only in the twentieth century, however, with the development of biochemistry, genetics, and molecular biology, that the full picture has emerged. Already in 1913 the distinguished Harvard biochemist Lawrence Henderson wrote: "The properties of matter and the course of cosmic evolution are now seen to be intimately related to the structure of the living being and to its activities; . . . the biologist may now rightly regard the Universe in its very essence as biocentric."[3] Henderson was led to this surprising view from his work on the regulation of acidity and alkalinity in living organisms, and the way that such regulation depends crucially upon the rather special properties of certain chemical substances. He was also greatly impressed at how water, which has a number of anomalous properties, is incorporated into life at a basic level. Had these various substances not existed, or had the laws of physics been somewhat different so that the substances did not enjoy these special properties, then life (at least as we know it) would be impossible. Henderson regarded the "fitness of the environment" for life as too great to be accidental, and asked what manner of law is capable of explaining such a match.

In the 1960s the astronomer Fred Hoyle noted that the element carbon, whose peculiar chemical properties make it crucial to terrestrial life, is manufactured from helium inside large stars. It is released therefrom by supernovae explosions, as discussed in the previous section. While investigating the nuclear reactions that lead to the formation of carbon in the stellar cores, Hoyle was struck by the fact that the key reaction proceeds only because of a lucky fluke. Carbon nuclei are made by a rather tricky process involving the simultaneous encounter of three high-speed helium nuclei, which then stick together. Because of the rarity of triple-nucleus encounters, the reaction can proceed at a significant rate only at certain well-defined energies (termed "resonances"), where the reaction rate is substantially amplified by quantum effects. By good fortune, one of these resonances is positioned just about right to correspond to the sort of energies that helium nuclei have inside large stars. Curiously, Hoyle did not know this at the time, but he predicted that it must be so on the basis that carbon is an abundant element in nature. Experiment subsequently proved him right. A detailed study also revealed other "coincidences" without which carbon would not be both produced and preserved inside stars. Hoyle was so impressed by this "monstrous series of accidents," he was prompted to comment that it was as if "the laws of nuclear physics have been deliberately designed with regard to the consequences they produce inside the stars."[4] Later he was to expound the view that the universe looks like a "put-up job," as though somebody had been "monkeying" with the laws of physics.[5]

These examples are intended merely as a sample. A long list of additional "lucky accidents" and "coincidences" has been compiled since, most notably by the astrophysicists Brandon Carter, Bernard Carr, and Martin Rees. Taken together, they provide impressive evidence that life as we know it depends very sensitively on the form of the laws of physics, and on some seemingly fortuitous accidents in the actual values that nature has chosen for various particle masses, force strengths, and so on. As these examples have been thoroughly discussed elsewhere, I will not list them here. Suffice it to say that, if we could play God, and select values for these quantities at whim by twiddling a set of knobs, we would find that almost all knob settings would render the universe uninhabitable. In some cases it seems as if the different knobs have to be fine-tuned to enormous precision if the universe is to be such that life will flourish. In their book *Cosmic Coincidences* John Gribbin and Martin Rees conclude: "The conditions in our Universe really do seem to be uniquely suitable for life forms like ourselves."[6]

It is a truism that we can only observe a universe that is consistent with our own existence. As I have mentioned, this linkage between human observership and the laws and conditions of the universe has become known, somewhat unfortunately, as the Anthropic Principle. In the trivial form just stated, the Anthropic Principle does not assert that our existence somehow *compels* the laws of physics to have the form they do, nor need one conclude that the laws have been deliberately designed with people in mind. On the other hand, the fact that even slight changes to the way things are might render the universe unobservable is surely a fact of deep significance.

Has the Universe Been Designed by an Intelligent Creator?

The early Greek philosophers recognized that the order and harmony of the cosmos demanded explanation, but the idea that these qualities derive from a creator working to a preconceived plan was well formulated only in the Christian era. In the thirteenth century, Aquinas offered the view that natural bodies act as if guided toward a definite goal or end "so as to obtain the best result." This fitting of means to ends implies, argued Aquinas, an intention. But, seeing as natural bodies lack consciousness, they cannot supply that intention themselves. "Therefore some intelligent being exists by whom all natural things are directed to their end; and this being we call God."[7]

Aquinas' argument collapsed in the seventeenth century with the development of the science of mechanics. Newton's laws explain the motion of material bodies perfectly adequately in terms of inertia and forces without the need for divine supervision. Nor did this purely mechanistic account of the world have any place for teleology (final or goal-directed causes). The explanation for the behavior of objects is to be sought in proximate physical causes—i.e., forces impressed upon them locally by other bodies. Nevertheless, this shift of world view did not entirely put paid to the idea that the world must have been designed for a purpose. Newton himself, as we have seen, believed that the solar system appeared too contrived to have arisen solely from the action of blind forces: "This most beautiful system of the sun, planets and comets, could only proceed from the counsel and dominion of an intelligent and powerful Being."[8] Thus, even within a mechanistic world view, one could still puzzle over the way in which material bodies have been arranged in the universe. For many scientists it was too much to suppose that the subtle and harmonious organization of nature is the result of mere chance.

This point of view was articulated by Robert Boyle, of Boyle's-law fame:

> The excellent contrivance of that great system of the world, and especially the curious fabric of the bodies of the animals and the uses of their sensories and other parts, have been made the great motives that in all ages and nations induced philosophers to acknowledge a Deity as the author of these admirable structures.[9]

Boyle introduced the famous comparison between the universe and a clockwork mechanism, which was most eloquently elaborated by the theologian William Paley in the eighteenth century. Suppose, argued Paley, that you were "crossing a heath" and came upon a watch lying on the ground. On inspecting the watch, you observed the intricate organization of its parts and how they were arranged together in a cooperative way to achieve a collective end. Even if you had never seen a watch and had no idea of its function, you would still be led to conclude from your inspection that this was a contrivance designed for a purpose. Paley then went on to argue that, when we consider the much more elaborate contrivances of nature, we should reach the same conclusion even more forcefully.

The weakness of this argument, exposed by Hume, is that it proceeds by analogy. The mechanistic universe is analogous to the watch; the watch had a designer, so therefore the universe must have had a designer. One might as well say that the universe is like an organism, so therefore it must have grown from a fetus in a cosmic womb! Clearly no analogical argument can amount to a proof. The best it can do is to offer support for a hypothesis. The degree of support will depend on how persuasive you find the analogy to be. As John Leslie points out, if the world were littered with pieces of granite stamped MADE BY GOD, after the fashion of the watchmaker's mark, surely even the Humes of this world should be convinced? "It can be asked whether every conceivable piece of seeming evidence of divine creative activity, including, say, messages written in the structures of naturally occurring chain molecules . . . would be shrugged off with the comment, 'Nothing improbable in that!' "[10] It is conceivable that clear evidence for design exists in nature, but that it is hidden in some way from us. Perhaps we will only become aware of the "architect's trademark" when we achieve a certain level of scientific attainment. This is the theme of the novel *Contact* by the astronomer Carl Sagan, in which a message is subtly embedded in the digits of pi—a number that is incorporated into the very structure of the universe—and only accessible by the use of sophisticated computer analysis.

It is also the case that most reasonable people accept other analogical arguments about the world. One example concerns the very existence of a physical world. Our immediate experiences always refer to our mental world, a world of sensory impressions. We usually think of this mental world as being a reasonably faithful map or model of a really existing physical world "out there," and distinguish between dream images and physical images. Yet a map or a model is also just an analogy; in this case it is one that we are usually prepared to accept. An even greater leap of faith is required when we conclude that there exist other minds besides our own. Our experience of other human beings derives entirely from interactions with their bodies: we cannot perceive their minds directly. Certainly other people behave *as if* they share our own mental experiences, but we can never know that. The conclusion that other minds exist is based entirely on analogy with our own behavior and experiences.

The design argument can't be categorized as right or wrong, but merely suggestive to a greater or lesser degree. So how suggestive is it? No scientist would today concur with Newton and claim that the solar system is too propitiously arranged to arise naturally. Although the origin of the solar system is not well understood, mechanisms are known to exist that could arrange the planets in the orderly manner that we find them. Nevertheless, the overall organization of the universe has suggested to many a modern astronomer an element of design. Thus James Jeans, who proclaimed that "the universe appears to have been designed by a pure mathematician" and it "begins to look more like a great thought than like a great machine," also wrote:

We discover that the universe shows evidence of a designing or controlling power that has something in common with our own individual minds—not, so far as we have discovered, emotion, morality, or aesthetic appreciation, but the tendency to think in the way which, for want of a better word, we describe as mathematical.[11]

Let's move on from astronomy for a moment. The most striking examples of "the contrivances of nature" are to be found in the biological domain, and it is to these that Paley devoted much of his attention. In biology the adaptation of means to ends is legendary. Consider the eye, for example. It is hard to imagine that this organ is not meant to provide the faculty of sight. Or that the wings of a bird aren't there for the purpose of flight. To Paley and many others, such intricate and successful adaptation bespoke providential arrangement by an intelligent designer. Alas, we all know about the speedy demise of this argument. Darwin's theory of evolution demonstrated decisively that complex organization efficiently adapted to the environment could arise as a result of random mutations and natural selection. No designer is needed to produce an eye or a wing. Such organs appear as a result of perfectly ordinary natural processes. A triumphalist celebration of this put-down is brilliantly presented in *The Blind Watchmaker* by the Oxford biologist Richard Dawkins.

The severe mauling meted out to the design argument by Hume, Darwin, and others resulted in its being more or less completely abandoned by theologians. It is all the more curious, therefore, that it has been resurrected in recent years by a number of scientists. In its new form the argument is directed not to the material objects of the universe as such, but to the underlying laws, where it is immune from Darwinian attack. To see why, let me first explain the essential character of Darwinian evolution. At its heart, Darwin's theory requires the existence of an ensemble, or a collection of similar individuals, upon which selection may act. For example, consider how polar bears may have come to blend so well with snow. Imagine a collection of brown bears hunting for food in snowy terrain. Their prey easily sees them coming and beats a hasty retreat. The brown bears have a hard time. Then, by some genetic accident, a brown bear gives birth to a white bear. The white bear makes a good living because it can creep up on its prey without being noticed so easily. It lives longer than its brown competitors and produces more white offspring. They too fare better, and produce still more white bears. Before long, the white bears are predominating, taking all the food, and driving the brown bears to extinction.

It's hard to imagine that something like the foregoing story isn't close to the truth. But notice how crucial it is that there be many bears to start with. One member of the bear ensemble is accidentally born white, and a selective advantage is gained over the others. The whole argument depends on nature being able to select from a collection of similar, competing individuals. When it comes to the laws of physics and the initial cosmological conditions, however, there is no ensemble of competitors. The laws and initial conditions are unique to our universe. (I shall come to the question of whether there may exist an ensemble of universes with differing laws shortly.) If it is the case that the existence of life requires the laws of physics and the initial conditions of the universe to be fine-tuned to high precision, and that fine-tuning does in fact obtain, then the suggestion of design seems compelling.

But before we leap to that conclusion, it is as well to consider some objections. First, it is sometimes argued that, if nature did not oblige by producing the right conditions for life to form, we would not ourselves be here to argue about the matter. That is of course true, but it hardly amounts to a counterargument. The fact is, we *are* here, and here by grace of some pretty felicitous arrangements. Our existence cannot of itself explain these arrangements. One could shrug the matter aside with the comment that we are certainly very lucky that the universe just happened to possess the necessary conditions for life to flourish, but that this is a meaningless quirk of fate. Again, it is a question of personal judgment. Suppose it could be demonstrated that life would be impossible unless the ratio of the mass of the electron to that of the proton was within 0.00000000001 percent of some completely independent number—say, one hundred times the ratio of the densities of water and mercury at 18 degrees centigrade (64.4 degrees Fahrenheit). Even the most hard-nosed skeptic must surely be tempted to conclude that there was "something going on."

So how are we to judge just how "fishy" the setup is? The problem is that there is no natural way to quantify the intrinsic improbability of the known "coincidences." From what range might the value of, say, the strength of the nuclear force (which fixes the position of the Hoyle resonances, for example) be selected? If the range is infinite, then any finite range of values might be considered to have zero probability of being selected. But then we should be equally surprised however weakly the requirements for life constrain those values. This is surely a *reductio ad absurdum* of the whole argument. What is needed is a sort of metatheory—a theory of theories—that supplies a well-defined probability for any given range of parameter values. No such metatheory is available, or has to my knowledge even been proposed. Until it is, the degree of "fishiness" involved must remain entirely subjective. Nevertheless, fishy it is!

Another objection sometimes raised is that life evolves to suit the prevailing conditions, so that it is no surprise after all to find life so well adapted to its circumstances. This may be true as far as the general state of the environment is concerned. Moderate climatic changes, for example, are likely to be accommodated. It would certainly be a mistake to point to the Earth and say: "Look how favorable the conditions are for life! The climate is just right, there is a plentiful supply of oxygen and water, the strength of gravity is just right for the size of limbs, etc., etc. What an extraordinary series of coincidences!" The Earth is but one planet among a huge ensemble spread throughout our galaxy and beyond. Life can form only on those planets where the conditions are suitable. Had the Earth not been one of them, then this book might have been written in another galaxy instead. We are not concerned here with anything so parochial as life on Earth. The question is, under what conditions might life arise at least somewhere in the universe? If that life arises, it will inevitably find itself situated in a suitable location.

The specialness argument I have been discussing refers not to this or that niche, but to the underlying laws of physics themselves. Unless those laws meet certain requirements, life won't even get started. Obviously carbon-based life could not exist if there were no carbon. But what about alternative life forms, much beloved of science-fiction writers? Once more we cannot really know. If the laws of physics differed a bit from their actual form, new possibilities for life might arise to replace the lost possibility of life as we know it. Set against that is the general view that biological mechanisms are actually rather specific and difficult to operate, and would be unlikely to emerge from a haphazard arrangement of physics. But until we have a proper understanding of the origin of life, or knowledge about alternative life forms elsewhere in the universe, the question must remain open.

The Ingenuity of Nature

To return once more to Einstein's famous utterance that "God is subtle but not malicious," we gain a clue to another intriguing aspect of the natural order. Einstein meant that to achieve an understanding of nature one must exercise considerable mathematical skill, physical insight, and mental ingenuity, but that nevertheless the goal of understanding is attainable. This is a topic I discussed in somewhat different language in chapter 6, where I pointed out that the world seems to be structured in such a way that its mathematical description is not at all trivial yet is still within the capabilities of human reasoning.

As I have remarked once or twice already, it is very difficult to communicate the concept of nature's mathematical subtlety to those unacquainted with mathematical physics, yet to the scientists involved what I am referring to is clear enough. It is, perhaps, at its most striking in the topics of particle physics and field theory, where several branches of advanced mathematics must be amalgamated. Put at its most crude: You find that a straightforward application of mathematics gets you so far, and then you get stuck. Some internal inconsistency appears, or else the theory yields results that are hopelessly at variance with the real world. Then some clever person comes along and discovers a mathematical trick—some obscure loophole in a theorem, perhaps, or an elegant reformulation of the original problem in entirely new mathematical language—and, hey presto, everything falls into place! It is impossible to resist the urge to proclaim nature at least as clever as the scientist for "spotting" the trick and exploiting it. One often hears theoretical physicists, speaking in the highly informal and colloquial way that they do, promoting their particular theory with the quip that it is so clever/subtle/elegant it is hard to imagine nature not taking advantage of it!

Let me give a thumbnail sketch of one example. In chapter 7 I discussed the recent attempts at the unification of the four fundamental forces of nature. Why should nature deploy four different forces? Wouldn't it be simpler, more efficient, and more elegant to have three, or maybe two or even one force, but with four distinct aspects? Or so it seemed to the physicists concerned, and so they looked for similarities between the forces to see if any mathematical amalgamation were possible. In the 1960s promising candidates were the electromagnetic force and the weak nuclear force. The electromagnetic force was known to operate through the exchange of particles called "photons." These photons flit back and forth between electrically charged particles such as electrons, and produce forces on them. When you rub a balloon and stick it to the ceiling, or feel the pull and push of magnets, you are witnessing this network of itinerant photons invisibly doing their work. You can think of these photons as rather like messengers, conveying the news about the force between particles of matter, which must then respond to it.

Now, theorists believed that something similar was going on inside nuclei when the weak nuclear force acts. A hypothetical particle, cryptically known as W, was invented to play a messenger role analogous to that of the photon. But whereas photons were familiar in the lab, nobody had ever seen a W, so the main guide in this theory was mathematics. The theory was recast in a way that brought out its essential similarity to electromagnetism most suggestively. The idea was that, if you have two mathematical schemes more or less the same, you can join them together and make do with a single, amalgamated scheme. Part of this rejigging meant introducing an additional messenger particle, known as Z, which resembles the photon even more closely than does W. The trouble was, even in this new improved mathematical framework the two schemes—electromagnetism and the weak-force theories—still differed in one rather basic way. Although Z and the photon share many properties, their masses have to be at opposite ends of the spectrum. This is because the mass of the messenger particle is related in a simple way to the range of the force: the more massive the messenger particle, the shorter the range of the corresponding force. Now, electromagnetism is a force of unlimited range, requiring a messenger particle of zero mass, whereas the weak force is confined to subnuclear distances and requires its messenger particles to be so massive they would outweigh most atoms.

Let me say a few words about the masslessness of the photon. The mass of a particle is related to its inertia. The smaller the mass, the smaller the inertia, so the faster it will accelerate when pushed. If a body has a very low mass, a given impulse will impart a very great speed. If you imagine particles with less and less mass, then their speeds will be greater and greater. You might think that a particle with zero mass would move at an infinite speed, but that is not so. The theory of relativity forbids travel faster than light, so zero-mass particles travel at the speed of light. Photons, being "particles of light," are the obvious example. By contrast, W and Z particles were predicted to have masses of about eighty and ninety times the mass of the proton (the heaviest known stable particle), respectively.

The problem the theorists faced in the 1960s was how to combine two elegant mathematical schemes describing the electromagnetic and weak forces if they differ so markedly in one important detail. The breakthrough came in 1967. Building on the mathematical framework constructed some time earlier by Sheldon Glashow, two theoretical physicists, Abdus Salam and Steven Weinberg, independently spotted a way forward. The essential idea was this. Suppose the great mass of W and Z is not a primary quality, but something acquired as a result of interaction with something else; that is, suppose these particles are not, so to speak, born massive, they are just carrying someone else's load? The distinction is a subtle one, but crucial. It means that the mass is attributed not to the underlying laws of physics, but to the particular *state* that the W and Z are usually found in.

An analogy might make the point clearer. Stand a pencil on its tip and hold it vertical. Now let go. The pencil will topple over and line up in some direction. Say it points northeast. The pencil reached that state as a result of the action of the Earth's gravity. But its "northeast-liness" is not an intrinsic quality of gravity. The Earth's gravity certainly has an intrinsic "up-down-ness," but not a north-south-ness or an east-west-ness or anything in between. Gravity makes no distinction between different horizontal directions. So the northeastness of the pencil is merely an incidental property of the system pencil-plus-gravity that reflects the particular *state* that the pencil happens to be in.

In the case of W and Z, the role of gravity is played by a hypothetical new field, called a Higgs field after Peter Higgs of the University of Edinburgh. The Higgs field interacts with W and Z and causes them to "topple over" in a symbolic sense. Rather than picking up "northeast-ness," they pick up mass—and lots of it. The way is now open to unification with the electromagnetic force, because, underneath, W and Z are "really" massless, like the photon. The two mathematical schemes can then be amalgamated, yielding a unified description of a single "electroweak" force.

The rest, as they say, is history. In the early 1980s accelerators at the Centre for European Nuclear Research (CERN), near Geneva, finally produced W particles, and then Z. The theory was brilliantly confirmed. Two of nature's forces were seen to be in reality two facets of a single force. The point I want to make is that nature has evidently spotted the loophole in the argument that you can't join massless and massive particles together. You can if you use the Higgs mechanism.

There is a postscript to this story. The Higgs field, which does the all-important job, has its own associated particle, called the "Higgs boson." It is probably very massive indeed, which means you need a lot of energy to make it. Nobody has yet detected a Higgs boson, but it is number one on the list of discoveries waiting to happen. Its production will be one of the main goals for the giant new accelerator planned for the Texas scrubland in the late nineties. Known as the "superconducting supercollider" (SSC), this monstrous machine will be some fifty miles in circumference, and will accelerate protons and antiprotons to unprecedented energies. Counterrotating beams will be allowed to collide, producing encounters of awesome ferocity. The hope is that the SSC will pack enough punch to make the Higgs boson. But the Americans will be racing the Europeans, who hope that a Higgs will show up in one of the machines at CERN. Of course, until one is found, we can't be sure that nature really is using the Higgs mechanism. Perhaps she has found an even cleverer way. Watch out for the final act in this drama!

A Place for Everything and Everything in Its Place

When scientists ask of their subject matter, "Why would nature bother with this?" or "What is the point of that?" they seem to be ascribing intelligent reasoning to nature. Though they usually intend such questions to be in a lighthearted spirit, there is a serious content too. Experience has shown that nature does share our sense of economy, efficiency, beauty, and mathematical subtlety, and this approach to research can often pay dividends (as with the unification of the weak and electromagnetic forces). Most physicists believe that beneath the complexities of their subject lies an elegant and powerful unity, and that progress can be made by spotting the mathematical "tricks" that nature has exploited to generate an interestingly diverse and complex universe from this underlying simplicity.

There is, for example, an unstated but more or less universal feeling among physicists that everything that exists in nature must have a "place" or a role as part of some wider scheme, that nature should not indulge in profligacy by manifesting gratuitous entities, that nature should not be arbitrary. Each facet of physical reality should link in with the others in a "natural" and logical way. Thus, when the particle known as the muon was discovered in 1937, the physicist Isidor Rabi was astonished. "Who ordered that?" he exclaimed. The muon is a particle more or less identical to the electron in all respects except its mass, which is 206.8 times bigger. This big brother to the electron is unstable, and decays after a microsecond or two, so it is not a permanent feature of matter. Nevertheless, it seems to be an elementary particle in its own right and not a composite of other particles. Rabi's reaction is typical. What is the muon for? Why does nature need another sort of electron, especially one that disappears so promptly. How would the world be different if the muon simply did not exist?

The problem has since become even more marked. There are now known to be *two* bigger brothers. The second, discovered in 1974, is called the "tauon." Worse still, other particles also possess highly unstable big brothers. The so-called quarks—the building blocks of nuclear matter, like protons and neutrons—each have two heavier versions too. There are also three varieties of neutrino. The situation is set out schematically in table 1. It seems that all known particles of matter can be arranged into three "generations." In the first generation are the electron, the electron-neutrino, and the two quarks called "up" and "down," which together build protons and neutrons. The particles in this first generation are all essentially stable, and go to make up the

TABLE 1

	LEPTONS	QUARKS
First generation	electron	down
	electron-neutrino	up
Second generation	muon	strange
	muon-neutrino	charmed
Third generation	tauon	bottom
	tauon-neutrino	top

The known particles of matter are composed of twelve basic entities. Six of these, called "leptons," are relatively light and interact only weakly. The remaining six, called "quarks," are heavy and strongly interacting, and make up nuclear matter. The particles can be arranged into three generations with similar properties.

ordinary matter of the universe that we see. The atoms of your body, and of the sun and stars, are composed of these first-generation particles.

The second generation seems to be little more than a duplicate of the first. Here one finds the muon, which so astonished Rabi. These particles (with the possible exception of the neutrino) are unstable, and soon decay into first-generation particles. Then, lo and behold, nature does it again, and gives us yet another replication of the pattern in generation three! Now, you may be wondering whether there is any end to this replication. Perhaps there is an infinity of generations, and what we are witnessing is really some simple repeating pattern. Most physicists disagree. In 1989 the new particle-accelerator at CERN, called Lep (for Large Electron-Position ring), was used to examine carefully the decay of the Z particle. Now, Z decays into neutrinos, and the decay rate depends on the number of distinct neutrino species available in nature, so a careful measurement of the rate can be used to deduce the number of neutrinos. The answer came out to be three, which suggests that there are just three generations.

So we have a puzzle: why three? One or infinity would be "natural," but three seems plain perverse. This "generation puzzle" has been the spur to some important theoretical work. The most satisfactory progress in particle physics has come from the use of a branch of mathematics known as "group theory." This is closely connected with the subject of symmetry, one of nature's "favorite" manifestations. Group theory can be used to connect apparently distinct particles into unified families. Now, there are definite mathematical rules about how these groups can be represented and combined, and how many of each type of particle they describe. The hope is that a group-theory description will emerge that commends itself on other grounds, but which will also demand three generations of particles. Nature's apparent profligacy will then be seen as a necessary consequence of some deeper unifying symmetry.

Of course, until that deeper unification is demonstrated, the generation problem seems to offer a counterexample to the argument that nature is subtly economical rather than maliciously arbitrary. But so confident am I that nature shares our sense of economy, I am happy to offer as a hostage to fortune that the generation problem will be solved within the next decade or so, and that its solution will provide further striking evidence that nature really does abide by the rule "A place for everything and everything in its place."

There is an interesting corollary to this generation game which reinforces my point. I have not been completely truthful in the entries to table 1. At the time of writing, the top quark has not yet been definitively identified. On several occasions it has been "discovered," only to be undiscovered shortly afterward. Now, you might wonder why physicists are so confident that the top quark exists that they are prepared to expend a significant fraction of their scarce resources looking for it. Suppose it doesn't exist? Suppose there is actually a gap in the table (which is, after all, a human construction), so that there aren't three generations at all, but two and three-quarters? Well, you'd be hard-pressed to find a physicist who really believed that nature would be so perverse, and when the top quark is discovered (as I have no doubt it eventually will be), it will provide another example of nature doing things tidily.

The generation problem is actually part of the larger unification scheme to which I have alluded, and which is being tackled head-on by a small army of theorists. John Polkinghorne, who used to be a particle physicist before he took up the priesthood, writes about the confidence that physicists have in this next step of the unification program:

> My erstwhile colleagues are labouring away in the endeavour to produce a theory yet more all-embracing. . . . I would say that at present their efforts have an air of contrivance, even desperation, about them. Some vital fact or idea seems still to be missing. However, I do not doubt that in due course some deeper understanding will be achieved and a more profound pattern discerned at the basis of physical reality.[12]

As I have mentioned, the so-called superstring theory is the current fashion, but doubtless something else will come along soon. Although major difficulties lie ahead, I agree with Polkinghorne. I cannot believe that these problems are truly insoluble, and that particle physics cannot be unified. All the pointers compel one to suppose that there is unity rather than arbitrariness beneath it all, despite the head-scratching.

As a final remark on the question of the "need" for all these particles, it is a curious thought that muons, though absent from ordinary matter, do play a rather important role in nature after all. Most of the cosmic rays that reach the surface of the Earth are in fact muons. These rays form part of the natural background of radiation, and contribute to the genetic mutations that drive evolutionary change. Therefore, at least to a limited extent, one can find use for muons in biology. This provides another example of the felicitous dovetailing of the large and the small that I mentioned earlier in this chapter.

Is There Need for a Designer?

I hope the foregoing discussion will have convinced the reader that the natural world is not just any old concoction of entities and forces, but a marvelously ingenious and unified mathematical scheme. Now, words like "ingenious" and "clever" are undeniably human qualities, yet one cannot help attributing them to nature too. Is this just another example of our projecting onto nature our own categories of thought, or does it represent a genuinely intrinsic quality of the world?

We have come a long way from Paley's watch. To return to my favorite analogy once more, the world of particle physics is more like a crossword than a clockwork mechanism. Each new discovery is a clue, which finds its solution in some new mathematical linkage. As the discoveries mount up, so more and more cross-links are "filled in," and one begins to see a pattern emerge. At present there remain many blanks on the crossword, but something of its subtlety and consistency can be glimpsed. Unlike mechanisms, which can slowly evolve to more complex or organized forms over time, the "crossword" of particle physics comes ready-made. The links do not evolve, they are simply there, in the underlying laws. We must either accept them as truly amazing brute facts, or seek a deeper explanation.

According to Christian tradition, this deeper explanation is that God has designed nature with considerable ingenuity and skill, and that the enterprise of particle physics is uncovering part of this design. If one were to accept that, the next question is: to what purpose has God produced this design? In seeking to answer the question, we need to take into account the many "coincidences" mentioned earlier in connection with the Anthropic Principle and the requirements of biological organisms. The apparent "fine-tuning" of the laws of nature necessary if conscious life is to evolve in the universe then carries the clear implication that God has designed the universe so as to permit such life and consciousness to emerge. It would mean that our own existence in the universe formed a central part of God's plan.

But does design necessarily imply a designer? John Leslie has argued that it doesn't. Recall that in his theory of creation the universe exists as a result of "ethical requirement." He writes: "A world existing as a result of an ethical need could be just the same, just as rich in seeming evidence of a designer's touch, whether or not the need depended for its influence on creative acts directed by a benevolent intelligence."[13] In short, a good universe would look designed to us, even if it had not been.

In *The Cosmic Blueprint*, I wrote that the universe looks *as if* it is unfolding according to some plan or blueprint. The idea is (partially) captured in a schematic way by figure 12, where the role of the blueprint (or cosmic computer program, if you prefer) is played by the laws of physics, and represented by the sausage machine. The input is the cosmic initial conditions, and the output is organized complexity, or depth. A variant on the image is shown in figure 13, where the input is matter and the output mind. The essential feature is that something of *value* emerges as the result of processing according to some ingenious pre-existing set of rules. These rules look *as if* they are the product of intelligent design. I do not see how that can be denied. Whether you wish to believe that they really *have* been so designed, and if so by what sort of being, must remain a matter of personal taste. My own inclination is to suppose that qualities such as ingenuity, economy, beauty, and so on have a genuine transcendent reality—they are not merely the product of human experience—and that these qualities are reflected in the structure of the natural world. Whether such qualities can themselves bring the universe into existence I don't know. If they could, one

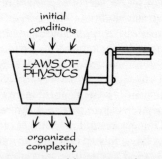

FIGURE 12. Symbolic representation of the cosmic evolution. The universe starts out in some relatively simple and featureless initial state, which is then "processed" by the fixed laws of physics to produce an output state which is rich in organized complexity.

Designer Universe

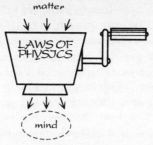

FIGURE 13. The evolution of matter from simplicity to complexity represented in figure 12 includes the production of conscious organisms from initially inanimate matter.

could conceive of God as merely a mythical personification of such creative qualities, rather than as an independent agent. This would, of course, be unlikely to satisfy anyone who feels he or she has a personal relationship with God.

Multiple Realities

Undoubtedly the most serious challenge to the design argument comes from the alternative hypothesis of many universes, or multiple realities. I introduced this theory in chapter 7 in connection with the cosmological argument for the existence of God. The basic idea is that the universe we see is but one among a vast ensemble. When deployed as an attack on the design argument, the theory proposes that all possible physical conditions are represented somewhere among the ensemble, and that the reason why our own particular universe looks designed is that only in those universes which have that seemingly contrived form will life (and hence consciousness) be able to arise. Hence it is no surprise that we find ourselves in a universe so propitiously suited to biological requirements. It has been "anthropically selected."

First we must ask what evidence there is for these other worlds. The philosopher George Gale has compiled a list of several physical theories that in one way or another imply an ensemble of universes. [14] The most frequently discussed theory of multiple universes concerns an interpretation of quantum mechanics. To see how quantum uncertainty leads to the possibility of more than one world, consider a simple example. Imagine a single electron immersed in a magnetic field. The electron possesses an intrinsic spin which endows it with a "magnetic moment." There will be an energy of interaction of the electron's magnetism with the external magnetic field, and this energy will depend on the angle between the direction of the imposed field and the direction of the electron's own magnetic field. If the fields are aligned, the energy will be low; if they are opposed, it will be high; and at intermediate angles it will vary between these values. We can effectively measure the orientation of the electron by making a measurement of this magnetic-interaction energy. What is found, and what is fundamental to the rules of quantum mechanics, is that only *two* values of the energy are ever observed, corresponding, roughly speaking, to the electron's magnetic moment pointing either along the magnetic field or opposed to it.

An interesting situation now arises if we set out deliberately to prepare the electron's magnetic field to be perpendicular to that of the imposed field. That is, we satisfy ourselves that the electron is pointing neither up nor down the external field, but across it. Mathematically, this arrangement is described by representing the electron by a state that is a "superposition" of the two possibilities. That is to say, the state is—again, roughly speaking—a hybrid of two overlapping realities: spin-up and spin-down. If, now, a measurement is made of the energy, the result will always be found as *either* up *or* down, and not some weird mixture of the two. But the inherent uncertainty of quantum mechanics forbids your knowing in advance which of these two possibilities will actually prevail. The rules of quantum mechanics will, however, allow you to assign *relative probabilities* to the alternatives. In the example considered there is equal probability for up or down. Then, according to a crude version of the many-universes theory, when a measurement is made, the universe splits into two copies, one in which the spin is up, the other in which it is down.

A more refined version envisages that there are always two universes involved, but that prior to the experiment they are identical in all respects. The effect of the experiment is to bring about their differentiation in respect of the electron's spin direction. In the case that the probabilities are unequal, one can imagine that there are many identical worlds in proportion to the relative probability. For example, if the probabilities were ⅔ up and ⅓ down, one could imagine three initially identical universes, two of which remain identical and have spin-up, the other differentiating itself by having spin-down. In general, one would need an infinite number of universes to cover all possibilities.

Now imagine extending this idea from a single electron to every quantum particle in the universe. Throughout the cosmos, the inherent uncertainties that confront each and every quantum particle are continually being resolved by differentiation of reality into ever more independently existing universes. This image implies that everything that can happen, will happen. That is, every set of circumstances that is physically possible (though not everything that is logically possible) will be manifested somewhere among this infinite set of universes.

The various universes must be considered to be in some sense "parallel" or coexisting realities. Any given observer will, of course, see only one of them, but we must suppose that the conscious states of the observer will be part of the differentiation process, so that each of the many alternative worlds will carry copies of the minds of the observers. It is part of the theory that you can't detect this mental "splitting"; each copy of us feels unique and integral. Nevertheless, there are stupendously many copies of ourselves in existence! Bizarre though the theory may seem, it is supported, in one version or another, by a large number of physicists as well as some philosophers. Its virtues are particularly compelling to those engaged in quantum cosmology, where alternative interpretations of quantum mechanics seem even less satisfactory. It must be said, however, that the theory is not without its critics, some of whom (e.g., Roger Penrose) challenge the claim that we would not notice the splitting.

This is by no means the only conjecture for an ensemble of worlds. Another, somewhat easier to visualize, is that what we have been calling "the universe" might just be a small patch of a much larger system extended in space. If we could look beyond the ten billion or so light-years accessible to our instruments, we would see (so the theory goes) other regions of the universe that are very different from ours. There is no limit to the number of different domains that could be included in this way, as the universe might be infinitely large. Strictly speaking, if we define "universe" to be everything that there is, then this is a many-regions rather than a many-universes theory, but the distinction is irrelevant for our purposes.

The question we now have to address is whether the evidence for design can equally well be taken as evidence for many universes. In some respects the answer is undoubtedly yes. For example, the spatial organization of the cosmos on a large scale is important for life. If the universe were highly irregular, it might produce black holes, or turbulent gas rather than well-ordered galaxies containing life-encouraging stable stars and planets. If you imagine a limitless variety of worlds in which matter was distributed at random, chaos would generally prevail. But here and there, purely by chance, an oasis of order would arise, permitting life to form. An adaptation of the inflationary-universe scenario along these lines has been proposed and studied by the Soviet physicist Andrei Linde. Although the quiescent oases would be almost unthinkably rare, it is no surprise that we find ourselves inhabiting one, for we could not live elsewhere. We are, after all, not surprised that we find ourselves atypically located on the surface of a planet, when the overwhelming proportion of the universe consists of near-empty space. So the cosmic order need not be attributed to the providential arrangement of things, but, rather, to the inevitable selection effect connected with our own existence.

This type of explanation might even be extended to some of the "coincidences" of particle physics. I discussed how the Higgs mechanism is invoked to explain the way in which the W and Z particles acquire their masses. In more elaborate unification theories, other Higgs fields are introduced to generate masses for all particles, and also to fix some other parameters of the theory related to force strengths. Now, just as in the falling-pencil analogy that I used on page 208, the system could topple into one of a whole variety of states (northeast, southeast, south-southwest, etc.), so, in these more elaborate Higgs-type mechanisms, the particle system can "topple" into different states. Which states are adopted would depend, randomly, on quantum fluctuations—i.e., on the inherent uncertainty built into quantum mechanics. In the many-universes theory one must suppose that every possible choice is somewhere represented by a complete universe. Alternatively, different choices might occur in different regions of space. Either way one would be presented with an ensemble of cosmological systems in which masses and forces took on different values. It would then be possible to argue that only where those quantities assumed the "coincidental" values needed for life, would life form.

In spite of the power of the many-universes theory to account for what would otherwise be considered remarkably special facts about nature, the theory faces a number of serious objections. The first of these I have already discussed in chapter 7, which is that it flies in the face of Occam's razor, by introducing vast (indeed infinite) complexity to explain the regularities of just one universe. I find this "blunderbuss" approach to explaining the specialness of our universe scientifically questionable. There is also the obvious problem that the theory can explain only those aspects of nature that are relevant to the existence of conscious life; otherwise there is no selection mechanism. Many of the examples I have given for design, such as the ingenuity and unity of particle physics, have little obvious connection with biology. Remember that it is not sufficient for the feature concerned simply to be relevant to biology, it has to be crucial to its actual prevalence.

Another point which is often glossed over is that, in all of the many-universe theories that derive from real physics (as opposed to simply fantasizing about the existence of other worlds), the laws of physics are the same in all the worlds. The selection of universes on offer is restricted to those that are *physically* possible, as opposed to those that can be imagined. There will be many more universes that are logically possible but contradict the laws of physics. In the example of the electron which can have either spin-up or -down, both worlds contain an electron with the same electric charge, obeying the same laws of electromagnetism, etc. So, although such many-universe theories might provide a selection of alternative *states* of the world, they cannot provide a selection of *laws*. It is true that the distinction between features of nature that owe their existence to a true underlying law, and those that can be attributed to the choice of state, is not always clear. As we have seen, certain parameters, like some particle masses, that previously were fixed into the theory as part of the assumed laws of physics, are now ascribed to states via the Higgs mechanism. But this mechanism can only work in a theory equipped with its own set of laws, and these will contain further features in need of explanation. Moreover, although quantum fluctuations might cause the Higgs mechanism to operate differently in different universes, it is far from clear in the theories being formulated at present that all possible values of particle masses, force strengths, etc., could be attained. Mostly the Higgs mechanism and similar so-called symmetry-breaking devices produce a discrete—indeed, finite—set of alternatives.

It is therefore not possible, as some physicists have suggested, to account for nature's *lawfulness* this way. Might it not be possible, however, to extend the many-universes idea to encompass different laws too? There is no logical objection to this, although there is no scientific justification for it either. But suppose one does entertain the existence of an even vaster stack of alternative realities for which any notion of law, order, or regularity is absent. Here chaos rules totally. The behavior of these worlds is entirely random. Well, just as a monkey tinkering with a typewriter will eventually type Shakespeare, so somewhere among that vast stack of realities will be worlds that are partially ordered, just by chance. Anthropic reasoning then leads us to conclude that any given observer will perceive an ordered world, mindbogglingly rare though such a world may be relative to its chaotic competitors. Would this account for our world?

I think the answer is clearly no. Let me repeat that anthropic arguments work only for aspects of nature that are crucial to life. If there is utter lawlessness, then the overwhelming number of randomly selected inhabited worlds will be ordered only in ways that are essential to the preservation of life. There is no reason, for example, why the charge of the electron need remain absolutely fixed, or why different electrons should have exactly the same charge. Minor fluctuations in the value of the electric charge would not be life-threatening. But what else keeps the value fixed—and fixed to such astonishing precision—if it is not a law of physics? One could, perhaps, imagine an ensemble of universes with a selection of laws, so that each universe comes with a complete and fixed set of laws. We could then perhaps use anthropic reasoning to explain why at least some of the laws we observe are what they are. But this theory must still presuppose the concept of law, and one can still ask where those laws come from, and how they "attach" themselves to universes in an "eternal" way.

My conclusion is that the many-universes theory can at best explain only a limited range of features, and then only if one appends some metaphysical assumptions that seem no less extravagant than design. In the end, Occam's razor compels me to put my money on design, but, as always in matters of metaphysics, the decision is largely a matter of taste rather than scientific judgment. It is worth noting, however, that it is perfectly consistent to believe in both an ensemble of universes and a designer God. Indeed, as I have discussed, plausible world-ensemble theories still require a measure of explanation, such as the lawlike character of the universes and why there exists a world-ensemble in the first place. I should also mention that discussions which start out with observations of only one universe and go on to make inferences about the improbability of this or that feature, raise some deep issues concerning the nature of probability theory. I believe that these have been satisfactorily attended to in John Leslie's treatment, but some commentators allege that attempts to argue backward "after the event"— the event in this case being our own existence—are fallacious.

Cosmological Darwinism

Recently an interesting adaptation of the many-universes theory has been proposed by Lee Smolin which avoids some of the objections to the other many-universes schemes by providing a curious linkage between the needs of living organisms and the multiplicity of the many universes. In chapter 2 I explained how the investigations of quantum cosmology suggest that "baby universes" can arise spontaneously as a result of quantum fluctuations, and that one may envisage a "mother universe" giving rise to progeny in this way. One circumstance under which new universes might be born is the formation of a black hole. According to classical (prequantum) gravitational theory, a black hole conceals a singularity, which can be regarded as a sort of edge of space-time. In the quantum version, the singularity is smeared out somehow. We don't know how, but it could be that the sharp boundary of space-time is replaced by a sort of tunnel or throat or umbilical cord connecting our universe to a new, baby universe. As explained in chapter 2, quantum effects would cause the black hole to evaporate eventually, severing the umbilical cord, and dispatching the baby universe to an independent career.

Smolin's refinement of this speculation is that the extreme conditions of near-singularity would have the effect of causing small random variations in the laws of physics. In particular, the values of some of the constants of nature, such as particle masses, charges, and so on, might be slightly different in the daughter universe from what they were in its mother. The daughter universe might then evolve slightly differently. Given enough generations, quite wide variations would occur among the many universes. It is likely, however, that those which differ substantially from our own would not evolve stars like ours (recall that the conditions for the formation of stars are rather special). Because black holes are most likely to form from dead stars, such universes would not produce many black holes, and hence would not give birth to many baby universes. By contrast, those universes with physical parameters well suited to form many stars would also form many black holes and thence many baby universes possessing similar values of these parameters. This difference in cosmic fecundity acts as a type of Darwinian selection effect. Although the universes don't actually compete, there are "successful" and "less successful" universes, so that the proportion of "successful" universes—in this case, efficient star-makers—in the total population will be rather large. Smolin then goes on to point out that the existence of stars is also an essential prerequisite for the formation of life. So the *same* conditions that encourage life also encourage the birth of other life-giving universes. In the Smolin scheme, life is not an extreme rarity, as it is in other many-universe theories. Instead, the large majority of universes are habitable.

In spite of its appeal, it is not clear that Smolin's theory makes any progress in explaining the specialness of the universe. The linkage between biological and cosmological selection is an attractive feature, but we can still wonder why the laws of nature are such that this linkage occurs. How fortunate that the requirements of life match those of the baby universes so well. Moreover, we still require the same basic structure of the laws in all these universes in order to make sense of the theory. That this basic structure also permits the formation of life remains a remarkable fact.

9

The Mystery at the End
of the Universe

"I have always thought it curious that, while most scientists
claim to eschew religion, it actually dominates their thoughts
more than it does the clergy."

Fred Hoyle

THE ESSENCE of this book has been to trace the logic of scientific
rationality back as far as it will go in the search for ultimate answers to
the mystery of existence. The idea that there might be a complete
explanation for everything—so that all of physical and metaphysical
existence would form a closed explanatory system—is a tantalizing one.
But what confidence can we have that the goal of this quest is not just
a chimera?

Turtle Power

In his famous book *A Brief History of Time* Stephen Hawking begins by
recounting a story about a woman who interrupts a lecture on the
universe to proclaim that she knows better. The world, she declares,
is really a flat plate resting on the back of a giant turtle. When asked
by the lecturer what the turtle rests on, she replied, "It's turtles all the
way down!"

The story symbolizes the essential problem that faces all who search
for ultimate answers to the mystery of physical existence. We would like
to explain the world in terms of something more fundamental, perhaps
a set of causes, which in turn rest upon some laws or physical principles,
but then we seek some explanation for this more fundamental level too,
and so on. Where can such a chain of reasoning end? It is hard to be
satisfied with an infinite regress. "No tower of turtles!" proclaims John

Wheeler. "No structure, no plan of organization, no framework of ideas underlaid by another structure or level of ideas, underlaid by yet another level, and yet another, *ad infinitum*, down to bottomless blackness."[1]

What is the alternative? Is there a "superturtle" that stands at the base of the tower, itself unsupported? Can this superturtle somehow "support itself"? Such a belief has a long history. We have seen how the philosopher Spinoza argued that the world could not have been otherwise, that God had no choice. Spinoza's universe is supported by the superturtle of pure logical necessity. Even those who believe in the contingency of the world often appeal to the same reasoning, by arguing that the world is explained by God, and that God is logically necessary. In chapter 7 I reviewed the problems that accompany these attempts to explain contingency in terms of necessity. The problems are no less severe for those who would abolish God and argue for some Theory of Everything that will explain the universe and will also be unique on the grounds of logical necessity.

It may seem as if the only alternatives are an infinite tower of turtles or the existence of an ultimate superturtle, the explanation for which lies within itself. But there is a third possibility: a closed loop. There is a delightful little book called *Vicious Circles and Infinity* which features a photograph of a ring of people (rather than turtles) each sitting on the lap of the person behind, and in turn supporting the one in front.[2] This closed loop of mutual support symbolizes John Wheeler's conception of the universe. "Physics gives rise to observer-participancy; observer-participancy gives rise to information; information gives rise to physics."[3] This rather cryptic statement is rooted in the ideas of quantum physics, where the observer and the observed world are closely interwoven: hence "observer-participancy." Wheeler's interpretation of quantum mechanics is that it is only through acts of observation that the physical reality of the world becomes actualized; yet this same physical world generates the observers that are responsible for concretizing its existence. Furthermore, this concretization extends even to the laws of physics themselves, for Wheeler rejects completely the notion of eternal laws: "The laws of physics cannot have existed from everlasting to everlasting. They must have come into being at the big bang."[4] So, rather than appeal to timeless transcendent laws to bring the universe into being, Wheeler prefers the image of a "self-excited circuit," wherein the physical universe bootstraps itself into existence, laws and all. Wheeler's own symbol for this closed-loop participatory universe is shown in figure 14. Neat though such "loopy" systems may be, they inevitably fall short of a complete explanation of things, for one can still ask "Why *that* loop?" or even "Why does *any* loop exist at all?" Even a closed loop of mutually-supportive turtles invites the question "Why turtles?"

All three of the above arrangements are founded on the assumption of human rationality: that it is legitimate to seek "explanations" for things, and that we truly understand something only when it is "explained." Yet it has to be admitted that our concept of rational explanation probably derives from our observations of the world and our evolutionary inheritance. Is it clear that this provides adequate guidance when we are tangling with ultimate questions? Might it not be the case that the reason for existence has no explanation in the usual sense? This does not mean that the universe is absurd or meaningless, only that an understanding of its existence and properties lies outside the usual categories of rational human thought. We have seen how application of human reasoning in its most refined and formalized sense—to mathematics—is nevertheless full of paradox and uncertainty. Gödel's theorem warns us that the axiomatic method of making logical deductions from given assumptions cannot in general provide a system which is both provably complete and consistent. There will always be truth that lies beyond, that cannot be reached from a finite collection of axioms.

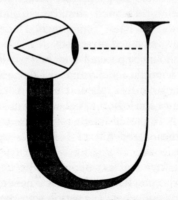

FIGURE 14. Symbolic representation of John Wheeler's "participatory universe." The large U stands for "universe," and the eye for observers who emerge at some stage, and then look back toward the origin.

The search for a closed logical scheme that provides a complete and self-consistent explanation for everything is doomed to failure. Like Chaitin's cabalistic number, such a thing may exist "out there" abstractly—indeed, its existence could be known to us, and we might get to know bits of it—but we cannot know its whole form on the basis of rational thought.

It seems to me that, as long as we insist on identifying "understanding" with "rational explanation" of the sort familiar in science, we will inevitably end up with turtle trouble: either an infinite regress, or a mysterious self-explaining superturtle, or an unexplained ring of turtles. There will always be mystery at the end of the universe. It may be, however, that there are other forms of understanding which will satisfy the inquiring mind. Can we make sense of the universe without turtle trouble? Is there a route to knowledge—even "ultimate knowledge"—that lies outside the road of rational scientific inquiry and logical reasoning? Many people claim there is. It is called mysticism.

Mystical Knowledge

Most scientists have a deep mistrust of mysticism. This is not surprising, as mystical thought lies at the opposite extreme to rational thought, which is the basis of the scientific method. Also, mysticism tends to be confused with the occult, the paranormal, and other fringe beliefs. In fact, many of the world's finest thinkers, including some notable scientists such as Einstein, Pauli, Schrödinger, Heisenberg, Eddington, and Jeans, have also espoused mysticism. My own feeling is that the scientific method should be pursued as far as it possibly can. Mysticism is no substitute for scientific inquiry and logical reasoning so long as this approach can be consistently applied. It is only in dealing with ultimate questions that science and logic may fail us. I am not saying that science and logic are likely to provide the wrong answers, but they may be incapable of addressing the sort of "why" (as opposed to "how") questions we want to ask.

The expression "mystical experience" is often used by religious people, or those who practice meditation. These experiences, which are undoubtedly real enough for the person who experiences them, are said to be hard to convey in words. Mystics frequently speak of an overwhelming sense of being at one with the universe or with God, of glimpsing a holistic vision of reality, or of being in the presence of a powerful and loving influence. Most important, mystics claim that they can grasp *ultimate reality* in a single experience, in contrast to the long and tortuous deductive sequence (petering out in turtle trouble) of the logical-scientific method of inquiry. Sometimes the mystical path seems to involve little more than an inner sense of peace—"a compassionate, joyful stillness that lies beyond the activity of busy minds" was the way a physicist colleague once described it to me. Einstein spoke of a "cosmic religious feeling" that inspired his reflections on the order and harmony of nature. Some scientists, most notably the physicists Brian Josephson and David Bohm, believe that regular mystical insights achieved by quiet meditative practices can be a useful guide in the formulation of scientific theories.

In other cases mystical experiences seem to be more direct and revelatory. Russell Stannard writes of the impression of facing an overpowering force of some kind, "of a nature to command respect and awe. . . . There is a sense of urgency about it; the power is volcanic, pent up, ready to be unleashed."[5] Science writer David Peat describes "a remarkable feeling of intensity that seems to flood the whole world around us with meaning. . . . We sense that we are touching something universal and perhaps eternal, so that the particular moment in time takes on a numinous character and seems to expand in time without limit. We sense that all boundaries between ourselves and the outer world vanish, for what we are experiencing lies beyond all categories and all attempts to be captured in logical thought."[6]

The language used to describe these experiences usually reflects the culture of the individual concerned. Western mystics tend to emphasize the personal quality of the presence, often describing themselves as being with someone, usually God, who is different from themselves but with whom a deep bond is felt. There is, of course, a long tradition of such religious experiences in the Christian Church and among the other Western religions. Eastern mystics emphasize the wholeness of existence and tend to identify themselves more closely with the presence. Writer Ken Wilber describes the Eastern mystical experience in characteristically cryptic language:

> In the mystical consciousness, Reality is apprehended directly and immediately, meaning without any mediation, any symbolic elaboration, any conceptualization, or any abstractions; subject and object become one in a timeless and spaceless act that is beyond any and all forms of mediation. Mystics universally speak of contacting reality in

227

its "suchness", its "isness", its "thatness", without any intermediaries; beyond words, symbols, names, thoughts, images.[7]

The essence of the mystical experience, then, is a type of shortcut to truth, a direct and unmediated contact with a perceived ultimate reality. According to Rudy Rucker:

> The central teaching of mysticism is this: *Reality is One.* The practice of mysticism consists in finding ways to experience this unity directly. The One has variously been called the Good, God, the Cosmos, the Mind, the Void, or (perhaps most neutrally) the Absolute. No door in the labyrinthine castle of science opens directly onto the Absolute. But if one understands the maze well enough, it is possible to jump out of the system and experience the Absolute for oneself. . . . But, ultimately, mystical knowledge is attained all at once or not at all. There is no gradual path. . . .[8]

In chapter 6 I described how some scientists and mathematicians claim to have had sudden revelatory insights akin to such mystical experiences. Roger Penrose describes mathematical inspirations as a sudden "breaking through" into a Platonic realm. Rucker reports that Kurt Gödel also spoke of the "other relation to reality," by which he could directly perceive mathematical objects, such as infinity. Gödel himself was apparently able to achieve this by adopting meditative practices, such as closing off the other senses and lying down in a quiet place. For other scientists the revelatory experience happens spontaneously, in the midst of the daily clamor. Fred Hoyle relates such an incident that occurred to him while he was driving through the North of England. "Rather as the revelation occurred to Paul on the Road to Damascus, mine occurred on the road over Bowes Moor." Hoyle and his collaborator Jayant Narlikar had, in the late 1960s, been working on a cosmological theory of electromagnetism that involved some daunting mathematics. One day, as they were struggling over a particularly complicated integral, Hoyle decided to take a vacation from Cambridge to join some colleagues hiking in the Scottish Highlands.

> As the miles slipped by I turned the quantum mechanical problem . . . over in my mind, in the hazy way I normally have in thinking mathematics in my head. Normally, I have to write things down on

paper, and then fiddle with the equations and integrals as best I can. But somewhere on Bowes Moor my awareness of the mathematics clarified, not a little, not even a lot, but as if a huge brilliant light had suddenly been switched on. How long did it take to become totally convinced that the problem was solved? Less than five seconds. It only remained to make sure that before the clarity faded I had enough of the essential steps stored safely in my recallable memory. It is indicative of the measure of certainty I felt that in the ensuing days I didn't trouble to commit anything to paper. When ten days or so later I returned to Cambridge I found it possible to write out the thing without difficulty.[9]

Hoyle also reports a conversation on the topic of revelation with Richard Feynman:

Some years ago I had a graphic description from Dick Feynman of what a moment of inspiration feels like, and of it being followed by an enormous sense of euphoria, lasting for maybe two or three days. I asked how often had it happened, to which Feynman replied 'four', at which we both agreed that twelve days of euphoria was not a great reward for a lifetime's work.[10]

I have recounted Hoyle's experience here rather than in chapter 6 because he himself describes it as a truly religious (as opposed to a merely Platonic) event. Hoyle believes that the organization of the cosmos is controlled by a "superintelligence" who guides its evolution through quantum processes, an idea I mentioned briefly in chapter 7. Furthermore, Hoyle's is a teleological God (somewhat like that of Aristotle or Teilhard de Chardin) directing the world toward a final state in the infinite future. Hoyle believes that by acting at the quantum level this superintelligence can implant thoughts or ideas from the future, ready-made, into the human brain. This, he suggests, is the origin of both mathematical and musical inspiration.

The Infinite

In our quest for ultimate answers it is hard not to be drawn, in one way or another, to the infinite. Whether it is an infinite tower of turtles, an infinity of parallel worlds, an infinite set of mathematical propositions, or an infinite Creator, physical existence surely cannot be rooted in anything finite. Western religions have a long tradition of identifying God with the Infinite, whereas Eastern philosophy seeks to eliminate the differences between the One and the Many, and to identify the Void and the Infinite—zero and infinity.

When the early Christian thinkers such as Plotinus proclaimed that God is infinite, they were primarily concerned to demonstrate that he is not limited in any way. The mathematical concept of infinity was at that time still fairly vague. It was generally believed that infinity is a limit toward which an enumeration may proceed, but which is unachievable in reality. Even Aquinas, who conceded God's infinite nature, was not prepared to accept that infinity had more than a potential, as opposed to an actual, existence. An omnipotent God "cannot make an absolutely unlimited thing," he maintained.

The belief that infinity was paradoxical and self-contradictory persisted until the nineteenth century. At this stage the mathematician Georg Cantor, while investigating problems of trigonometry, finally succeeded in providing a rigorous logical demonstration of the self-consistency of the actually infinite. Cantor had a rough ride with his peers, and was dismissed by some eminent mathematicians as a madman. In fact, he did suffer mental illness. But eventually the rules for the consistent manipulation of infinite numbers, though often strange and counterintuitive, came to be accepted. Indeed, much of twentieth-century mathematics is founded on the concept of the infinite (or infinitesimal).

If infinity can be grasped and manipulated using rational thought, does this open the way to an understanding of the ultimate explanation of things without the need for mysticism? No, it doesn't. To see why, we must take a look at the concept of infinity more closely.

One of the surprises of Cantor's work is that there is not just one infinity but a multiplicity of them. For example, the set of all integers and the set of all fractions are both infinite sets. One feels intuitively that there are more fractions than integers, but this is not so. On the other hand, the set of all decimals is bigger than the set of all fractions, or all integers. One can ask: is there a "biggest" infinity? Well, how about combining all infinite sets together into one superduperset? The class of all possible sets has been called Cantor's Absolute. There is one snag. This entity is not itself a set, for if it were it would by definition include itself. But self-referential sets run smack into Russell's paradox.

And here we encounter once more the Gödelian limits to rational thought—the mystery at the end of the universe. We cannot know Cantor's Absolute, or any other Absolute, by rational means, for any Absolute, being a Unity and hence complete within itself, must include itself. As Rucker remarks in connection with the Mindscape—the class of all sets of ideas—"If the Mindscape is a One, then it is a member of itself, and thus can only be known through a flash of mystical vision. No rational thought is a member of itself, so no rational thought could tie the Mindscape into a One."[11]

What Is Man?

"I do not feel like an alien in this universe."

Freeman Dyson

Does the frank admission of hopelessness discussed in the previous section mean that all metaphysical reasoning is valueless? Should we adopt the approach of the pragmatic atheist who is content to take the universe as given, and get on with cataloguing its properties? There is no doubt that many scientists are opposed temperamentally to any form of metaphysical, let alone mystical arguments. They are scornful of the notion that there might exist a God, or even an impersonal creative principle or ground of being that would underpin reality and render its contingent aspects less starkly arbitrary. Personally I do not share their scorn. Although many metaphysical and theistic theories seem contrived or childish, they are not obviously more absurd than the belief that the universe exists, and exists in the form it does, reasonlessly. It seems at least worth trying to construct a metaphysical theory that reduces some of the arbitrariness of the world. But in the end a rational explanation for the world in the sense of a closed and complete system of logical truths is almost certainly impossible. We are barred from ultimate knowledge, from ultimate explanation, by the very rules of reasoning that prompt us to seek such an explanation in the first place. If we wish to progress beyond, we have to embrace a different concept of "understanding" from that of rational explanation. Possibly the mystical path is a way to such an understanding. I have never had a mystical experience myself, but I keep an open mind about the value of such experiences. Maybe they provide the only route beyond the limits to which science and philosophy can take us, the only possible path to the Ultimate.

The central theme that I have explored in this book is that, through science, we human beings are able to grasp at least some of nature's secrets. We have cracked part of the cosmic code. Why this should be, just why *Homo sapiens* should carry the spark of rationality that provides the key to the universe, is a deep enigma. We, who are children of the universe—animated stardust—can nevertheless reflect on the nature of that same universe, even to the extent of glimpsing the rules on which it runs. How we have become linked into this cosmic dimension is a mystery. Yet the linkage cannot be denied.

What does it mean? What is Man that we might be party to such privilege? I cannot believe that our existence in this universe is a mere quirk of fate, an accident of history, an incidental blip in the great cosmic drama. Our involvement is too intimate. The physical species *Homo* may count for nothing, but the existence of mind in some organism on some planet in the universe is surely a fact of fundamental significance. Through conscious beings the universe has generated self-awareness. This can be no trivial detail, no minor byproduct of mindless, purposeless forces. We are truly meant to be here.

Notes

CHAPTER 1: Reason and Belief

1. "The Rediscovery of Time" by Ilya Prigogine, in *Science and Complexity* (ed. Sara Nash, Science Reviews Ltd, London, 1985), p. 23.
2. *God and Timelessness* by Nelson Pike (Routledge & Kegan Paul, London, 1970), p. 3.
3. *Trinity and Temporality* by John O'Donnell (Oxford University Press, Oxford, 1983), p. 46.

CHAPTER 2: Can the Universe Create Itself?

1. "The History of Science and the Idea of an Oscillating Universe" by Stanley Jaki, in *Cosmology, History and Theology* (eds. W. Yourgrau & A. D. Breck, Plenum, New York and London, 1977), p. 239.
2. *Confessions* by Augustine, book 12, ch. 7.
3. *Against Heresies* by Iranaeus, book III, X, 3.
4. "Making Sense of God's Time" by Russell Stannard, *The Times* (London), 22 August 1987.
5. *A Brief History of Time* by Stephen W. Hawking (Bantam, London and New York, 1988), p. 136.
6. Ibid., p. 141.
7. "Creation as a Quantum Process" by Chris Isham, in *Physics, Philosophy and Theology: A Common Quest for Understanding* (eds. Robert John Russell, William R. Stoeger, and George V. Coyne, Vatican Observatory, Vatican City State, 1988), p. 405.

Notes

8. "Beyond the Limitations of the Big Bang Theory: Cosmology and Theological Reflection" by Wim Drees, *Bulletin of the Center for Theology and the Natural Sciences* (Berkeley) 8, No. 1 (1988).

CHAPTER 3: What Are the Laws of Nature?

1. *Theories of Everything: The Quest for Ultimate Explanations* by John Barrow (Oxford University Press, Oxford, 1991), p. 6.
2. Ibid., p. 58.
3. "Discourse on metaphysics" by G. W. Leibniz, in *Philosophical Writings* (ed. G. H. R. Parkinson, Dent, London, 1984).
4. *Theories of Everything* by Barrow, p. 295.
5. *The Grand Titration: Science and Society in East and West* by Joseph Needham (Allen & Unwin, London, 1969).
6. *Theories of Everything* by Barrow, p. 35.
7. *The Cosmic Code* by Heinz Pagels (Bantam, New York, 1983), p. 156.
8. "Plato's Timaeus and Contemporary Cosmology: A Critical Analysis" by F. Walter Mayerstein, in *Foundations of Big Bang Cosmology* (ed. F. W. Mayerstein, World Scientific, Singapore, 1989), p. 193.
9. Reprinted in *Einstein: A Centenary Volume* (ed. A. P. French, Heinemann, London, 1979), p. 271.
10. "Rationality and Irrationality in Science: From Plato to Chaitin" by F. Walter Mayerstein, University of Barcelona report, 1989.
11. *Cosmic Code* by Pagels, p. 157.
12. "Excess Baggage" by James Hartle, in *Particle Physics and the Universe: Essays in Honour of Gell-Mann* (ed. J. Schwarz, Cambridge University Press, Cambridge, 1991), in the press.
13. "Singularities and Time-Asymmetry" by Roger Penrose, in *General Relativity: An Einstein Centenary Survey* (eds. S. W. Hawking and W. Israel, Cambridge University Press, Cambridge, 1979), p. 631.
14. "Excess Baggage" by Hartle, in *Particle Physics and the Universe* (in the press).

CHAPTER 4: Mathematics and Reality

1. *The Mathematical Principles of Natural Philosophy*, A. Motte's translation, revised by Florian Capori (University of California Press, Berkeley and Los Angeles, 1962), vol. 1, p. 6.
2. "Review of *Alan Turing: The Enigma*," reprinted in *Metamagical Themas* by Douglas Hofstadter (Basic Books, New York, 1985), p. 485.
3. "Quantum Theory, the Church-Turing Principle and the Universal Quan-

tum Computer" by David Deutsch, *Proceedings of the Royal Society London* A 400 (1985), p. 97.

4. "The Unreasonable Effectiveness of Mathematics" by R. W. Hamming, *American Mathematics Monthly* 87 (1980), p. 81.

5. *The Recursive Universe* by William Poundstone (Oxford University Press, Oxford, 1985).

6. "Artificial Life: A Conversation with Chris Langton and Doyne Farmer," *Edge* (ed. John Brockman, New York), September 1990, p. 5.

7. *Recursive Universe* by Poundstone, p. 226.

8. Quoted in *Recursive Universe* by Poundstone.

CHAPTER 5: Real Worlds and Virtual Worlds

1. "Computer Software in Science and Mathematics" by Stephen Wolfram, *Scientific American* 251 (September 1984), p. 151.

2. "Undecidability and Intractability in Theoretical Physics" by Stephen Wolfram, *Physical Review Letters* 54 (1985), p. 735.

3. "Computer Software" by Wolfram, p. 140.

4. "Physics and Computation" by Tommaso Toffoli, *International Journal of Theoretical Physics* 21 (1982), p. 165.

5. "Simulating Physics with Computers" by Richard Feynman, *International Journal of Theoretical Physics* 21 (1982), p. 469.

6. "The Omega Point as *Eschaton*: Answers to Pannenberg's Questions for Scientists" by Frank Tipler, *Zygon* 24 (1989), pp. 241–42.

7. *The Anthropic Cosmological Principle* by John D. Barrow and Frank J. Tipler (Oxford University Press, Oxford, 1986), p. 155.

8. "On Random and Hard-to-Describe Numbers" by Charles Bennett, IBM report 32272, reprinted in "Mathematical Games" by Martin Gardner, *Scientific American* 241 (November 1979), p. 31.

9. Ibid., pp. 30–1.

10. *Theories of Everything: The Quest for Ultimate Explanations* by John Barrow (Oxford University Press, Oxford, 1991), p. 11.

11. "Dissipation, Information, Computational Complexity and the Definition of Organization" by Charles Bennett, in *Emerging Syntheses in Science* (ed. D. Pines, Addison-Wesley, Boston, 1987), p. 297.

CHAPTER 6: The Mathematical Secret

1. "The Unreasonable Effectiveness of Mathematics in the Natural Sciences" by Eugene Wigner, *Communications in Pure and Applied Mathematics* 13 (1960), p. 1.

Notes

2. *Mathematics and Science* (ed. Ronald E. Mickens, World Scientific Press, Singapore, 1990).

3. *The Emperor's New Mind: Concerning Computers, Minds and the Laws of Physics* by Roger Penrose (Oxford University Press, Oxford, 1989), p. 111.

4. Ibid., p. 95.

5. Ibid.

6. Martin Gardner, foreword to ibid., p. vi.

7. Ibid., p. 97.

8. *Infinity and the Mind* by Rudy Rucker (Birkhauser, Boston, 1982), p. 36.

9. *Emperor's New Mind* by Penrose, p. 428.

10. *The Psychology of Invention in the Mathematical Field* by Jacques Hadamard (Princeton University Press, Princeton, 1949), p. 13.

11. Quoted in ibid., p. 12.

12. *Emperor's New Mind* by Penrose, p. 420.

13. Quoted in *Mathematics* by M. Kline (Oxford University Press, Oxford, 1980), p. 338.

14. Quoted in *Superstrings: A Theory of Everything?* by P. C. W. Davies and J. R. Brown (Cambridge University Press, Cambridge, 1988), pp. 207–8.

15. "Computation and Physics: Wheeler's Meaning Circuit?" by Rolf Landauer, *Foundations of Physics* 16 (1986), p. 551.

16. *Theories of Everything: The Quest for Ultimate Explanation* by John Barrow (Oxford University Press, Oxford, 1991), p. 172.

17. *Emperor's New Mind* by Penrose, p. 430.

CHAPTER 7: Why Is the World the Way It Is?

1. For the full quotation and a discussion of this point see *The World Within the World* by John D. Barrow (Oxford University Press, Oxford, 1990), p. 349.

2. Message of His Holiness Pope John Paul II, in *Physics, Philosophy and Theology: A Common Quest for Understanding* (eds. Robert John Russell, William R. Stoeger, and George V. Coyne, Vatican Observatory, Vatican City State, 1988), p. M1.

3. "No Faith in the Grand Theory" by Russell Stannard, *The Times* (London), 13 November 1989.

4. *Theories of Everything: The Quest for Ultimate Explanation* by John Barrow (Oxford University Press, Oxford, 1991), p. 210.

5. *Divine and Contingent Order* by Thomas Torrance (Oxford University Press, Oxford, 1981), p. 36.

6. *A Brief History of Time* by Stephen W. Hawking (Bantam, London and New York, 1988), p. 174.

7. "Excess Baggage" by James Hartle, in *Particle Physics and the Universe: Essays*

in Honour of Gell-Mann (ed. J. Schwarz, Cambridge University Press, Cambridge, 1991), in the press.

8. "Ways of Relating Science and Theology" by Ian Barbour, in *Physics, Philosophy and Theology* (eds. Russell et al.), p. 34.
9. *Brief History* by Hawking, p. 174.
10. *Divine and Contingent Order* by Torrance, pp. 21, 26.
11. *Science and Value* by John Leslie (Basil Blackwell, Oxford, 1989), p. 1.
12. *The World Within the World* by Barrow, p. 292.
13. Ibid., p. 349.
14. *Theories of Everything* by Barrow, p. 2.
15. *The Emperor's New Mind: Concerning Computers, Minds and the Laws of Physics* by Roger Penrose (Oxford University Press, Oxford, 1989), p. 421.
16. *Rational Theology and the Creativity of God* by Keith Ward (Pilgrim Press, New York, 1982), p. 73.
17. Ibid., p. 3.
18. Ibid., pp. 216–17.
19. *The Reality of God* by Schubert M. Ogden (SCM Press, London, 1967), p. 17.
20. "On Wheeler's Notion of 'Law Without Law' in Physics" by David Deutsch, in *Between Quantum and Cosmos: Studies and Essays in Honor of John Archibald Wheeler* (ed. Alwyn Van der Merwe et al., Princeton University Press, Princeton, 1988), p. 588.
21. *Theories of Everything* by Barrow, p. 203.
22. *Rational Theology* by Ward, p. 25.
23. "Argument from the Fine-Tuning of the Universe" by Richard Swinburne, in *Physical Cosmology and Philosophy* (ed. J. Leslie, Macmillan, London, 1990), p. 172.

CHAPTER 8: Designer Universe

1. *The First Three Minutes* by Steven Weinberg (Andre Deutsch, London, 1977), p. 149.
2. *Chance and Necessity* by Jacques Monod, trans. A. Wainhouse (Collins, London, 1972), p. 167.
3. *The Fitness of the Environment* by L. J. Henderson (reprinted by Peter Smith, Gloucester, Mass., 1970), p. 312.
4. Quoted in *Religion and the Scientists* (ed. Mervyn Stockwood, SCM, London, 1959), p. 82.
5. *The Intelligent Universe* by Fred Hoyle (Michael Joseph, London, 1983), p. 218.
6. *Cosmic Coincidences* by John Gribbin and Martin Rees (Bantam Books, New York and London, 1989), p. 269.

Notes

7. *Summa Theologiae* by Thomas Aquinas, pt. I, ques. II, art. 3.
8. *Philosophiae Naturalis Principia Mathematica* by Isaac Newton (1687), bk. III, General Scholium.
9. "A Disquisition About the Final Causes of Natural Things," in *Works* by Robert Boyle (London, 1744), vol. 4, p. 522.
10. *Universes* by John Leslie (Routledge, London and New York, 1989), p. 160.
11. *The Mysterious Universe* by James Jeans (Cambridge University Press, Cambridge, 1931), p. 137.
12. "The Faith of a Physicist" by John Polkinghorne, *Physics Education* 22 (1987), p. 12.
13. *Value and Existence* by John Leslie (Basil Blackwell, Oxford, 1979), p. 24.
14. "Cosmological Fecundity: Theories of Multiple Universes" by George Gale, in *Physical Cosmology and Philosophy* (ed. J. Leslie, Macmillan, London, 1990), p. 189.

CHAPTER 9: The Mystery at the End of the Universe

1. "Information, Physics, Quantum: The Search for Links" by John Wheeler, in *Complexity, Entropy and the Physics of Information* (ed. Wojciech H. Zurek, Addison-Wesley, Redwood City, California, 1990), p. 8. See also n. 21 to ch. 7.
2. *Vicious Circles and Infinity: An Anthology of Paradoxes* by Patrick Hughes and George Brecht (Doubleday, New York, 1975), Plate 15.
3. "Information" by Wheeler, p. 8.
4. Ibid., p. 9.
5. *Grounds for Reasonable Belief* by Russell Stannard (Scottish Academic Press, Edinburgh, 1989), p. 169.
6. *The Philosopher's Stone: The Sciences of Synchronicity and Creativity* by F. David Peat (Bantam Doubleday, New York, 1991), in the press.
7. *Quantum Questions* (ed. Ken Wilber, New Science Library, Shambhala, Boulder, and London, 1984), p. 7.
8. *Infinity and the Mind* by Rudy Rucker (Birkhauser, Boston, 1982), pp. 47, 170.
9. "The Universe: Past and Present Reflections" by Fred Hoyle, University of Cardiff report 70 (1981), p. 43.
10. Ibid., p. 42.
11. *Infinity* by Rucker, p. 48.

Select Bibliography

Barbour, Ian G. *Religion in an Age of Science* (SCM Press, London, 1990).

Barrow, John. *The World Within the World* (Clarendon Press, Oxford, 1988).

Barrow, John. *Theories of Everything: The Quest for Ultimate Explanation* (Oxford University Press, Oxford, 1991).

Barrow, John D., and Tipler, Frank J. *The Anthropic Cosmological Principle* (Clarendon Press, Oxford, 1986).

Birch, Charles. *On Purpose* (New South Wales University Press, Kensington, 1990).

Bohm, David. *Wholeness and the Implicate Order* (Routledge & Kegan Paul, London, 1980).

Coveney, Peter, and Highfield, Roger. *The Arrow of Time* (W. H. Allen, London, 1990).

Craig, William Lane. *The Cosmological Argument from Plato to Leibniz* (Macmillan, London, 1980).

Drees, Wim B. *Beyond the Big Bang: Quantum Cosmologies and God* (Open Court, La Salle, Illinois, 1990).

Dyson, Freeman. *Disturbing the Universe* (Harper & Row, New York, 1979).

Ferris, Timothy. *Coming of Age in the Milky Way* (Morrow, New York, 1988).

French, A. P., ed. *Einstein: A Centenary Volume* (Heinemann, London, 1979).

Gleick, James. *Chaos: Making a New Science* (Viking, New York, 1987).

Harrison, Edward R. *Cosmology* (Cambridge University Press, Cambridge, 1981).

Hawking, Stephen W. *A Brief History of Time* (Bantam, London and New York, 1988).

Langton, Christopher G., ed. *Artificial Life* (Addison-Wesley, Reading, Mass., 1989).

Leslie, John. *Value and Existence* (Basil Blackwell, Oxford, 1979).

Select Bibliography

Leslie, John. *Universes* (Routledge, London and New York, 1989).

Leslie, John, ed. *Physical Cosmology and Philosophy* (Macmillan, London, 1990).

Lovell, Bernard. *Man's Relation to the Universe* (Freeman, New York, 1975).

MacKay, Donald M. *The Clockwork Image* (Inter-Varsity Press, London, 1974).

McPherson, Thomas. *The Argument from Design* (Macmillan, London, 1972).

Mickens, Ronald E., ed. *Mathematics and Science* (World Scientific Press, Singapore, 1990).

Monod, Jacques. *Chance and Necessity*, trans. A. Wainhouse (Collins, London, 1972).

Morris, Richard. *Time's Arrows* (Simon and Schuster, New York, 1984).

Morris, Richard. *The Edges of Science* (Prentice-Hall Press, New York, 1990).

Pagels, Heinz. *The Dreams of Reason* (Simon and Schuster, New York, 1988).

Pais, Abraham. *Subtle Is the Lord: The Science and the Life of Albert Einstein* (Clarendon Press, Oxford, 1982).

Peacocke, A. R., ed. *The Sciences and Theology in the Twentieth Century* (Oriel, Stocksfield, England, 1981).

Penrose, Roger. *The Emperor's New Mind: Concerning Computers, Minds and the Laws of Physics* (Oxford University Press, Oxford, 1989).

Pike, Nelson. *God and Timelessness* (Routledge & Kegan Paul, London, 1970).

Poundstone, William. *The Recursive Universe* (Oxford University Press, Oxford, 1985).

Prigogine, Ilya, and Stengers, Isabelle. *Order Out of Chaos* (Heinemann, London 1984).

Rowe, William. *The Cosmological Argument* (Princeton University Press, Princeton, 1975).

Rucker, Rudy. *Infinity and the Mind* (Birkhauser, Boston, 1982).

Russell, Robert John; Stoeger, William R.; and Coyne, George V., eds. *Physics, Philosophy and Theology: A Common Quest for Understanding* (Vatican Observatory, Vatican City State, 1988).

Silesius, Angelus. *The Book of Angelus Silesius*, trans. Frederick Franck (Vintage Books, New York, 1976).

Silk, Joseph. *The Big Bang* (Freeman, New York, 1980).

Stannard, Russell. *Grounds for Reasonable Belief* (Scottish Academic Press, Edinburgh, 1989).

Swinburne, Richard. *The Coherence of Theism* (Clarendon Press, Oxford, 1977).

Torrance, Thomas. *Divine and Contingent Order* (Oxford University Press, Oxford, 1981).

Trusted, Jennifer. *Physics and Metaphysics: Facts and Faith* (Routledge, London, 1991).

Ward, Keith. *Rational Theology and the Creativity of God* (Pilgrim Press, New York, 1982).

Ward, Keith. *The Turn of the Tide* (BBC Publications, London, 1986).

Select Bibliography

Weinberg, Steven. *The First Three Minutes* (Andre Deutsch, London, 1977).

Wilber, Ken, ed. *Quantum Questions* (New Science Library, Shambhala, Boulder, and London, 1984).

Zurek, Wojciech H., ed. *Complexity, Entropy and the Physics of Information* (Addison-Wesley, Redwood City, California, 1990).

Index

Index

Index

in artificial intelligence, 32, 125
artificial life and, 24, 115–16, 124–26,
    187–88
basic operation of, 113
cellular automata and, 110–16, 119,
    121–22, 123, 124, 125, 136, 138,
    166, 168
complexity and, 14, 111–16, 128–39,
    213
essence of, 118
halting problem in, 106–7, 133–34
heat generated by, 120, 121, 122
input-output data of, 97, 113, 118,
    119, 120–22, 123, 129–30, 133,
    135, 136, 137, 214
languages of, 131
logic gates of, 113–14, 120–21
programs of, 97, 99, 112, 116, 118–
    120, 122–23, 126, 128–30, 131,
    133–34, 135–39, 187–88, 214
reality simulations by, 118–28, 187–88
time reversibility and, 120–23
universal, 99, 103–7, 108, 110, 112,
    113, 114, 119, 131
universe as, 22, 97, 99, 123–26, 135–
    139, 146–48, 214–15
see also machines
constants of nature, 158, 221
Contact (Sagan), 202
contingent order, 69, 163–65, 166, 167,
    169–72, 177, 178–81, 182–85, 188–
    190, 191, 224
Conway, John, 110–11, 113–14, 115–
    116, 125
Copernicus, Nicolaus, 20, 45
Cosmic Blueprint, The (Davies), 78, 136,
    169, 183, 191, 192, 214
cosmic code, 78–80, 132, 137, 148–50,
    155, 174, 232
Cosmic Code, The (Pagels), 79
Cosmic Coincidences (Gribbin and Rees),
    200
cosmic rays, 212–13
cosmological argument, 39–40, 42, 172–
    185, 188–89, 193, 215
cosmology, 13, 20, 31, 32, 36, 38, 39–
    72, 147–48, 166, 182, 196
    initial conditions in, 14, 63, 88–92,
    97, 116, 158–60, 168, 169, 173,
    193, 195, 204, 214

inflationary, 70, 89, 90, 91, 197, 218
oscillating model in, 50–55
quantum, 14, 55, 61–69, 70–72, 73,
    90–91, 92, 147–48, 158–60, 168,
    169, 217, 221
participatory universe in, 224–25, 226
steady-state model in, 55–57, 61,
    170–71
teleological, 36, 74, 182, 200, 229
time in, 40–41, 45–55, 57, 61, 62–69,
    70–72, 170–71
time-reversing model in, 53
time-symmetric model in, 53–55
see also big bang; God; many universes,
    theory of
cosmos, 43
creation, see cosmology; big bang
creation ex nihilo, doctrine of, 36, 39–40,
    41–42, 44–45, 56, 68–69, 161, 171,
    179–81
creation field, 56, 57
cyclic time, 40–41, 50–55

dark matter, 51
Darwin, Charles, 20, 115, 153, 203, 222
Dawkins, Richard, 203
Dead of Night (film), 55
deism, 43, 58
Demiurge, 35–36, 43, 44, 79, 80–81,
    141, 178
depth, logical, 137–39, 196, 214
Descartes, René, 34, 76, 77, 140, 166,
    178
design, in the universe, 194–222
    biology in, 20, 21, 24, 31, 32, 195,
    198–200, 203–5, 213, 215, 217,
    218–19, 220, 221, 222, 231–32
    God and, 200–205, 206, 213–15, 220
    ingenuity of, 206–9, 213, 219
    many-universes theory and, 215–22
    neatness of, 209–13
    unity in, 156, 195–98, 209, 212, 213,
    219, 222
determinism, 29–31, 111, 114, 122, 182,
    192
Deutsch, David, 107–8, 186–87
Devi, Shakuntala, 154
Diophantine equations, 132–33
Dirac, Paul, 149, 176
dissipative machines, 120–21, 122

245

Index

Index

Index

inflationary universe, theory of, 70, 89, 90, 91, 197, 218
information, 85, 121, 128, 147, 174, 192
  algorithmic, 129–35, 137
  bits of, 118, 129, 130
initial conditions, 87–92, 122, 123, 135, 138
  in cosmology, 14, 63, 88–92, 97, 116, 158–60, 168, 169, 173, 193, 195, 204, 214
input-output data, computer, 97, 113, 118, 119, 120–22, 123, 129–30, 133, 135, 136, 137, 214
inspiration, 28, 145, 228–29
intelligence, 43–44, 116, 200–205, 229
  alien, 24, 151
  artificial, 32, 125
  see also brain; mind
Iranaeus, 45
Isham, Chris, 68
Islam, 45, 75, 134, 191
"Is the End in Sight for Theoretical Physics?" (Hawking), 86, 166

Jaki, Stanley, 40–41
Jeans, James, 140, 151, 202–3, 226
John Paul II, Pope, 164
Josephson, Brian, 227
Judaism, 36, 41–42, 75
  Cabalist, 95, 134, 225

Kant, Immanuel, 23, 31–32, 126, 150, 186
Kempthorn, 75–76
Kepler, Johannes, 76, 95
Kolmogorov, Andrei, 129
Kowa Seki, 77
Kruskal universe model, 33

Landauer, Rolf, 121, 146–48
Langton, Christopher G., 115
Laplace, Pierre, 97, 182
latitude, 96
Leibniz, G. W., 76–77, 171, 172–73
Lep, 210
leptons, 211
Leslie, John, 171, 172, 201–2, 214, 221
life, see biology

Life, game of, 110–11, 112, 113–16, 119, 121, 122, 125, 136, 138
light, 46, 47–48, 57, 82, 157
  photons of, 206–7, 208
  speed of, 48, 135, 147, 150, 207
lightning calculators, 154
Linde, Andrei, 70, 218
linearity, 78, 157–58, 159, 182
living organism, universe as, 22, 36, 74, 151, 182, 201
locality, 157, 158–59
logic, 19, 22, 25–26, 27–28, 32, 72, 86, 101, 103, 115, 121, 133, 134, 162, 165–66, 169, 172, 185, 193, 225
  premises of, 25, 26, 166, 188
  mysticism vs., 226–29
  quantum, 26
logical depth, 137–39, 196, 214
logical necessity, 161–62, 165, 172, 178, 185, 186, 187–88, 195, 224
logic gates, computer, 113–14, 120–21
longitude, 96

machines:
  calculating, 98, 100, 104, 106, 107
  clockwork, universe as, 22, 29, 42, 43, 76, 97, 111, 115, 181, 182, 200–201, 213
  dissipative, 120–21, 122
  self-reproducing, 111–16, 118, 168
  Turing, 99, 103–7, 109, 112, 113, 114, 122
  see also computers
magnetic moment, 216
Mandelbrot, Benoit, 142
Mandelbrot set, 142–43
many universes, theory of, 126, 190–91, 204, 215–22
  mother and child universes in, 70–72, 221–22
  objections to, 218–21
  spatial regions in, 217, 218
Margolus, Norman, 114, 122
mass, 207–9, 210, 218, 219, 221
mathematics, 15, 19, 22, 31, 32, 72, 77, 79, 83, 84, 92, 93–160, 173–77, 186, 191–93, 206, 207, 209, 213, 216
  of ancient Greeks, 23, 26, 27, 29, 86–87, 93–95, 96, 108, 134, 140, 165

248

Index

READ MORE IN PENGUIN

In every corner of the world, on every subject under the sun, Penguin represents quality and variety – the very best in publishing today.

For complete information about books available from Penguin – including Puffins, Penguin Classics and Arkana – and how to order them, write to us at the appropriate address below. Please note that for copyright reasons the selection of books varies from country to country.

In the United Kingdom: Please write to *Dept. EP, Penguin Books Ltd, Bath Road, Harmondsworth, West Drayton, Middlesex UB7 ODA*

In the United States: Please write to *Consumer Sales, Penguin USA, P.O. Box 999, Dept. 17109, Bergenfield, New Jersey 07621-0120*. VISA and MasterCard holders call 1-800-253-6476 to order Penguin titles

In Canada: Please write to *Penguin Books Canada Ltd, 10 Alcorn Avenue, Suite 300, Toronto, Ontario M4V 3B2*

In Australia: Please write to *Penguin Books Australia Ltd, P.O. Box 257, Ringwood, Victoria 3134*

In New Zealand: Please write to *Penguin Books (NZ) Ltd, Private Bag 102902, North Shore Mail Centre, Auckland 10*

In India: Please write to *Penguin Books India Pvt Ltd, 706 Eros Apartments, 56 Nehru Place, New Delhi 110 019*

In the Netherlands: Please write to *Penguin Books Netherlands bv, Postbus 3507, NL-1001 AH Amsterdam*

In Germany: Please write to *Penguin Books Deutschland GmbH, Metzlerstrasse 26, 60594 Frankfurt am Main*

In Spain: Please write to *Penguin Books S. A., Bravo Murillo 19, 1° B, 28015 Madrid*

In Italy: Please write to *Penguin Italia s.r.l., Via Felice Casati 20, I–20124 Milano*

In France: Please write to *Penguin France S. A., 17 rue Lejeune, F–31000 Toulouse*

In Japan: Please write to *Penguin Books Japan, Ishikiribashi Building, 2–5–4, Suido, Bunkyo-ku, Tokyo 112*

In Greece: Please write to *Penguin Hellas Ltd, Dimocritou 3, GR–106 71 Athens*

In South Africa: Please write to *Longman Penguin Southern Africa (Pty) Ltd, Private Bag X08, Bertsham 2013*

BY THE SAME AUTHOR

About Time
Einstein's Unfinished Revolution

In this, his latest book, Paul Davies confronts the tough questions about time. He gives straightforward descriptions of topics such as the theory of relativity and Hawking's 'imaginary time' and concludes that the revolution begun by Einstein remains tantalizingly incomplete.

'A tour of some of the most exciting – and outlandish – work in modern physics . . . Writing with passion and wit, he lets his scientific message shine through' – Peter Tallack in the *New Statesman & Society*

Are We Alone?
Implications of the Discovery of Extraterrestrial Life

Since ancient times people have been fascinated by the idea of extraterrestrial life; today we are searching systematically for it. Paul Davies's striking new book examines the assumptions that go into the search and draws out the startling implications for science, religion and our world-view should we discover that we are not alone.

Superforce
Revised edition

Many scientists believe we are on the verge of a 'Theory of Everything' – a complete unification of all the fundamental forces and particles of nature. The search for a unifying superforce extends from the microworld within the atom to the large scale structure of the cosmos, weaving gravitation, electromagnetism and nuclear forces into a seamless fabric.

also published:

The Edge of Infinity   **Other Worlds**
The Cosmic Blueprint   **God and the New Physics**

and, with John Gribbin:

The Matter Myth